I0569029

OLD WEST PARANORMAL VOL. 1

TERROR IN THE MINES!

OLD WEST PARANORMAL VOL. 1

TERROR IN
THE MINES!

Noe Torres & John LeMay

DEAD HORSE HISTORY

A SUBSIDIARY OF BICEP BOOKS. ROSWELL, NEW MEXICO

Copyright © 2024 Noe Torres & John LeMay. Published by Dead Horse History, a subsidiary of Bicep Books, Roswell, New Mexico.

Dead Horse History logo by Melanie Hubner.

All rights reserved. No part of this publication may be reproduced, distributed, or transmitted in any form or by any means, including photocopying, recording, or other electronic or mechanical methods, without the prior written permission of the publisher, except in the case of brief quotations embodied in critical reviews and certain other noncommercial uses permitted by copyright law.

Printed in the United States of America

Torres, Noe and John LeMay.
Old West Paranormal Volume I: Terror in the Mines!
ISBN 978-1-953221-15-5
Mining—Ghosts/Supernatural

For the wonderful, visionary authors whose books enriched my childhood and sent our imaginations soaring, including Ray Bradbury, Isaac Asimov, Arthur C. Clarke, Robert Heinlein, Jules Verne, H. G. Wells, Sir Arthur Conan Doyle, and Edgar Rice Burroughs.

"Coal Mine Explosion" by Granger, 1873.

A NOTE FROM THE AUTHOR

S ince 2011, my friend and co-author John LeMay and I have written a series of books about unexplained phenomena that happened during America's "Old West" period, roughly from the mid 1800s to the year 1900. We did exhaustive historical research, using the newspapers and other documents of the time, to uncover stories of unusual sightings of strange creatures, bizarre flying objects, futuristic contraptions, and many other events that remain unexplained even today.

And, while writing our many volumes, we noticed a recurring theme – tales involving strange, unexplained events that happened in mines, or that involved miners, in the Old West. Stories of these strange mining-related phenomenon soon became a significant subset of the tales we included in our series of books about this fascinating period in American history.

It seems that miners, prospectors, and other similarly-engaged workers who wandered the rugged landscape of the West sometimes experienced some very frightening occurrences, often deep in the bowels of the Earth, as they worked to dig up buried treasures.

Over the years, John and I often spoke of collecting all the strange tales involving "mines and miners" into a single volume that finally lays out in detail everything we have learned about this bizarre phenomenon. The result is this book you now hold in your hands.

Perhaps there are secrets that still lie buried in the subterranean reaches of our planet. Strange creatures, unexplained artifacts, and objects that seem out of place in time and space. As we begin our journey into the mysteries and terrors of these Old West mines, let's keep in mind that even today, scientists are often befuddled by strange discoveries that are made during modern mining operations and other similar construction activities.

So together, with a torch in hand, let's carefully make our way past the foreboding entrance to this dark, deserted mine, and let's slowly and carefully work our way down toward the lower depths in hopes of finding new treasures, while hopefully keeping our sanity in the process.

Noe Torres

Edinburg, Texas
June 13, 2024

TABLE OF CONTENTS

INTRODUCTION
HAUNTED GOLD

G hosts and gold go together about as well as peanut butter and jelly. As it stands, ghosts often act as guardians of treasure caches. Such stories were especially prevalent in the era of the Old West, where prospectors spoke of gold-guarding ghosts and, occasionally, monstrous creatures like the dragons of old sitting atop treasure hordes.

Although mostly forgotten today, amongst the lost treasure lore of yore there was a strange subset of it that pertained to magic and witchcraft. For instance, one way of using witchcraft to guard one's buried gold was to murder a man at the spot of the treasure so that his ghost would guard it. Or, if not murdered there, a man who died at the spot of the treasure might become its guardian ghost. That's why some treasure hunters would sometimes make blood offerings at alleged treasure sites to appease the treasure guardian.

Among these mystical treasure hunters was none other than Mormon founder Joseph Smith, who implemented many rituals in his efforts. Because many regarded Smith as a "seer," or someone who could locate missing objects, he was occasionally hired by treasure seekers to find buried loot. To seek out said treasures, he would often use divining rods and peep stones. Smith would also draw magical circles around the dig site as protection against any potential treasure guardians. He even broke ground with a small silver shovel for good luck before transitioning to a normal sized shovel.

However, Smith claimed that when translating the Book of Mormon that he was commanded to no longer use his seeing abilities to search for earthly treasures, and so ceased the process. As stated before, Smith wasn't the only treasure hunter to use mystical means to look for gold, there were many others. And furthermore, it wasn't just Anglos who attributed supernatural elements to gold.

"Money Diggers" by John Quidor (1832).

The Apache afforded a supernatural respect to gold because it was the color of the sun, and thus they associated it with their one god, Ussen. As such, while one could pluck gold from the ground if the earth had already given it, to dig into the ground was forbidden. If one dug for gold, they ran the risk of upsetting the earth and an earthquake might result. A good example of this superstition was the story of the Lost Adams Diggings, a canyon full of gold in New Mexico Territory. It was discovered by a group of prospectors, led by a man named Adams, sometime during the Civil War.

The men were overjoyed when they found a stream running through the canyon that yielded many golden nuggets out in the deserts of Apacheria. All was well until a group of Apache riders led by Chief Nana stopped by their encampment. Nana instructed the men that they were welcome to take as much gold as they wanted so long as they didn't dig above the falls, for there resided the mother lode, or the main vein from which the gold in the river came. Naturally, the men went against the chief's wishes and began mining the gold from above the waterfall. Soon after, the Apache returned and massacred everyone but Adams, who escaped to tell the tale. However, in the years to follow, Adams could never again find the canyon. It was as if the landscape had changed drastically. In the mid-1880s, this was confirmed by a man named Jason Baxter. Baxter had himself found the same canyon in the 1870s and mounted a return expedition in the 1880s. However, when he returned to the canyon, he found that it had been caved in during a past cataclysm. The Apache were apparently right, and the spirits concealed the gold-bearing canyon in an earthquake.

Depiction of miners panning for gold.

J. Frank Dobie, who brought the Lost Adams story to the world in *Apache Gold & Yaqui Silver*, also touched on the mystical side of treasure hunting in his seminal work, *Coronado's*

Children. In it, Dobie revealed he possessed a copy of an archaic treasure hunting tome known as *The Sixth and Seventh Books of Moses,* published in 1880 but likely written much earlier. Of the book, Dobie called it "a curious and absurd farrago of matter pertaining to black magic and other superstitions…" that originated from Germany.[1]

THE

SIXTH AND SEVENTH

BOOKS OF MOSES;

OR,

MOSES' MAGICAL SPIRIT-ART,

KNOWN AS THE

WONDERFUL ARTS

OF THE OLD WISE HEBREWS, TAKEN FROM THE MOSAIC
BOOKS OF THE CABALA AND THE TALMUD,
FOR THE GOOD OF MANKIND.

Translated from the German, Word for Word, according to Old
Writings.

WITH NUMEROUS ENGRAVINGS.

New York:

1880.

Dobie elaborated that

> Instead of tabulating signs that indicate the way to treasure, the work affords directions for recovering it. Thus, Mephistopheles is an effective agent to "bring treasure from the earth and from the deep very quickly." The best time to enlist the aid of angels in the business of recovering treasure is in the sign of Scorpio, for then "they rule over legacies and riches." If devils and angels are lacking, certain cabalistic configurations will draw treasure upwards.[2]

Better known and possibly more ancient than *The Sixth and Seventh Books of Moses* was the German "Infernal Compulsion" manuscript. In 1715, a group of German treasure hunters used this forbidden text to summon spirits to help them locate a hidden cache of wealth rumored to be buried near their village of Jena. It had all started when a local tailor was looking over his vineyard in the moonlight. He sighted a ghostly woman in white pacing back and forth across his property. Unlike the Southwest's woman in white, La Llorona, who signifies death, in Germany, a woman in white was sometimes thought to signify a treasure spirit that could guide the way to untold riches. The tailor knew the dangers of following spirits to find treasure, and so sought out safety in numbers and acquired several partners to share in the danger, and possibly the wealth.

Among the party was a man in possession of the dreaded Infernal Compulsion manuscript, which was locked shut with two lead locks, each engraved with magical symbols. It was his idea that they should use the book to conjure the woman in white directly. They decided to do so on the night of Christmas Eve of 1715. The tailor was frightened of the idea, and decided to let the three partners go out on his land while he stayed safely locked within his home. So, the three went to a small cottage on the tailor's property and enacted the ritual. It did not go well.

The next morning, two of them were found expired, but not of cold or frostbite. It was as though they had simply dropped dead, and their bodies were covered with mysterious red markings. The occultist to whom the book belonged remembered beginning the ritual the night before but blacking out before he could repeat the required incantation. The next morning, he awoke inside the cottage alone and then went outside to look for his two companions, whom he found dead and covered in the red marks. The man turned himself into the authorities for the crime of communing with the dead. They arrested him and also took possession of the two dead bodies, locking them up in a separate cell. On duty in the jail that night were three guards—only one would survive. The surviving guard claimed that a dastardly ghost had appeared after midnight and killed his two companions. After that, the conjurer and the tailor alike were banished from Germany and the event was thereafter known as the Jena Christmas Eve Tragedy.

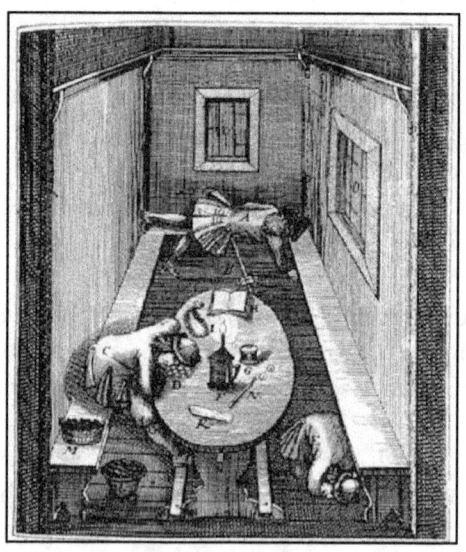

The story is notable because several of the tales in this volume are quite similar and told of groups of miners and amateur fortune seekers encountering ghostly spirits within mines and at spots said to harbor buried riches. Were these tales of terror in the mines true, or simply folktales carried over from Europe and transposed onto American soil? Read ahead and judge such stories for yourself…

Section Notes

---

[1] Dobie, *Coronado's Children*, p.302.
[2] Ibid, p.304.

SEARCHING FOR GOLD

WEIRD STORY OF TREASURE GUARD-
ED BY GHOSTS.

**Excitement Is Rife at Hot Springs, Ark.,
Over a Mysterious Chart—People Started
Digging, but Were Scared Off by Spirits.
Awaiting an Indian's Arrival.**

The people of South Hot Springs, a sub-
urb of Hot Springs, Ark., are greatly ex-
cited over the supposed existence of a hidden treasure in that locality. A few days
ago a stranger named Smith arrived there
and gave it out on the quiet that $222,000
worth of gold bullion was buried in the
vicinity many years ago by his grandfather and a Spaniard while fleeing from the
Indians. He was armed with a plat made
out in Spanish, which he claimed enabled
him to find the exact locality. Smith says
before his father's death he gave to him
the charts and told him about the place of
concealment.

Smith took into his confidence S. B.
Riles of Hot Springs, and the two went
in search of the spot. They soon encountered a thick patch of woods at the foot of
West mountain, and Smith, who is nearly
blind, was describing to his companion
the supposed whereabouts of the buried
treasure when suddenly he was overcome
by choking sensations and influenced so
visibly and unaccountably that he had to
be carried from the spot, and as soon as
the spell left him no amount of persuasion
could induce him to again go near the
place. He claimed that he had been there
several times before and alone to seek the
treasure, but the same horrible influence
prevented his operation every time.

Anderson Daily Bulletin (January 16, 1897).

1.

HAUNTED GOLD OF HOT SPRINGS

JANUARY 1897
SOUTH HOT SPRINGS, ARKANSAS

This first story began when a blind stranger named Smith drifted into South Hot Springs, Arkansas, whispering of a gold cache totaling $222,000 buried somewhere in the vicinity. The gold had been hidden by Smith's father and a companion from Spain some years ago. They had to bury it because they were being pursued by an unspecified Native American tribe. Smith's father told his son of the buried gold on his deathbed, and also produced for him some plats in Spanish said to lead to the site of the gold.

In Hot Springs, the blind Smith made a partner out of S.B. Riles. If Riles would help Smith get to the treasure, he would share in the wealth. Together the duo went to the spot indicated in the plat, which was simply called West Mountain at the time. There, in a thicket of woods, was the buried gold. According to the article, published in the *Anderson Daily Bulletin* of January 16, 1897, the location of the gold was as follows:

> Those marked trees are said to form a circle around the spot where the treasure is buried, and the large trees all have peculiar marks, which are described in the chart. The following is only a brief detailed description, as called for by the waybill and plat, and found exactly as

stated: One is to follow the circle of trees with seven hacks on them, and in that circle, and also similarly marked with seven hacks, are to be found the following large trees and their peculiar marks: On a white oak a certain number of feet from a hickory, leaning south, three scalp marks on opposite side; west of the hickory, on a sweet gum tree, the print of a hand on southeast side, south of which can be found a tree with a deer's foot on blaze, below which are the letters, "G. O.," the meaning of which, as told in the plat, is "gold ore;" east so many feet is a large flat rook, on the surface of which are three pig tracks; so many feet on each side of rock, maple trees, with rocks in fork, on which are hand or "whooper" marks. Many other trees and their peculiar marks are described exactly, besides which two iron stabs, the location of which was described by the plat, were found, and on them was peculiar lettering, as called for by the paper.

Postcard of Hot Springs, Arkansas.

However, upon reaching the spot, Smith was overtaken with a choking sensation and had to be carried away. The attack was a supernatural one, and Smith revealed that he had actually come to that spot in the past and had been attacked in a similar manner. In those days, Smith had apparently been able to see and had found the location on his own. Therefore, Smith was not only needing someone to guide him back to the spot, but also someone who the spirits hopefully would not attack.

Men constructing the brick archway over Hot Springs Creek.
(National Parks Service)

After the attack, Smith refused to return to the area. But, as the story spread, residents of Hot Springs decided to search it out for themselves. Meanwhile, Smith sent for an unnamed Native American friend who he believed could retrieve the treasure. This man, a chief of the Choctaw Nation, was apparently a descendant of the tribe who had chased Smith's grandfather and who could hopefully appease the spirits. As for the men digging around the tree grove, they soon experienced the same spiritual attacks suffered by Smith. According to the paper, a man identified as Tom Crouch saw what he called "an Indian ghost." His account in the paper was as follows:

Frank Johnson, myself and five other men started to work about 8 o'clock last night digging for the treasure. We had worked about an hour and had made good headway. Johnson was doing the digging at the time, when suddenly he dropped his pick, threw up his hands and fell back unconscious. I was standing close to him and caught him in my arms as he was falling, he regained consciousness in a few minutes, and I started toward home with him. When we reached the road some distance below, Johnson said he felt all right and could go home alone.

"I left him and returned to the digging. In a few minutes afterward we heard a noise on the mountain above us like a man walking on the rocks. We called, but received no answer. Finally we decided to investigate and see who it was or what it was, and found, to our utter surprise, that it was Johnson wandering aimlessly about in the dark. I spoke to him, but could get no response, and then I went up and caught hold of his hand and arm and shook him, and he appeared to suddenly arouse as if from a deep sleep. Then he told us what he had seen. He said he had seen a ghost, or spirit, and described it as an Indian about the size of four ordinary men, dressed in the regulation costume of the savage warrior, who, he said, had given vent to a genuine war-whoop at the time. None of the other members of our party had seen or heard anything of the kind, although a little later I saw a wild flash of lightning and felt a peculiar sensation come over me.

"I feel certain now that when Johnson returned to the spot after he had fainted he was under the influence of the spirit Indian, and that had we not interfered with him we would have been led to the identical spot where the treasure is buried."

Terror in the Mines!

The article continued that Mr. Crouch had spoken with others that had similar paranormal encounters at the dig-site. Unfortunately, no updates on the treasure were given, and the article concluded that

> Much excitement has been created among the denizens of South Hot Springs by the strange and startling experience of these men, as they related it, and the belief in the existence of the buried treasure is thereby strengthened among them. They are now awaiting the arrival of the old Indian chief from the Choctaw Nation.

If he ever arrived, and what he found if so, remains a mystery.

When the Workmen in an Old Mexican Mine Disturbed a Skeleton Found in a Chamber of the Mine, They Claim a Ghostly Figure, Clad in Old Spanish Armor, Sprang Upon Them, Sword in Hand.

Illustration from the *Chicago Saturday Blade* (May 11, 1907).

2.
A MINE POSSESSED
BY EVIL SPIRITS

In the January 9, 1903 edition of the *Alpine (Texas) Avalanche* newspaper, there appeared a strange news item about an abandoned silver mine in West Texas that seemed to be possessed by evil spirits. Interestingly, the story was first carried in mid-December of 1902 by several East Coast newspapers, including the *Saturday (Utica, N.Y.) Globe* and the *New York Sun*, and then was reprinted by other papers in the weeks that followed.

According to the story, the silver mine, called "Refugio" at this time but later called "Muerta," was located sixty miles from Alpine, Texas, near Chispa Mountain, whose name means "spark" in Spanish. Although the newspaper articles say that the mine is sixty miles *southwest* of Alpine, Chispa Mountain is actually *northwest* of Alpine.

The story goes that around 1882, American mining engineer Henry Boyd, who lived in San Antonio, Texas, was doing title research on Mexican mining properties at the Coahuila archives in Saltillo, Mexico, when he came across a mention of the abandoned silver mine in West Texas. The records indicated that the mine had been in operation from 1751 to 1791 and that it produced more than seven million dollars' worth of silver ore. The revenue from the mine went mostly to the government of Spain. After Texas gained its independence from Mexico in 1845, the property fell into the domain of the United States.

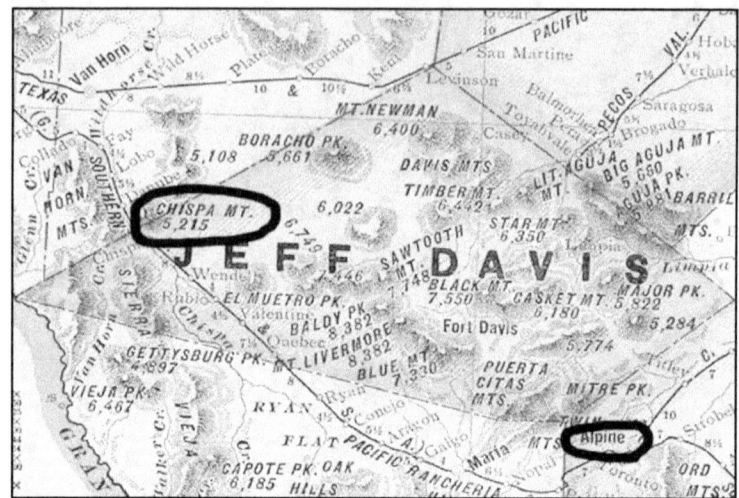

Circa 1920 map showing Chispa Mountain and Alpine.

According to the newspaper article, as soon as Boyd found out about the abandoned Spanish mine, he hired a Mexican guide named Pedro and immediately left Saltillo to go in search of the mine, whose location was close to 1,000 miles away in a desolate corner of Texas. The article said, "It was a long and fatiguing trip, and the two men experienced great hardships until they reached the little Mexican settlements along the Rio Grande south of Alpine, where they made their headquarters while they made expeditions into the rough country north of the river in search of the mine."

The men were having no luck finding the mine until a Mexican sheepherder told Boyd that he could lead them to the site of an abandoned smelter in a deep canyon. After a patient search, Boyd and his guide found the entrance to a vertical mine shaft and noticed a crude rope ladder dangling down into the opening.

With his guide remaining above ground, Boyd descended the ladder for about 100 feet, where he found the ancient mine workings, including a great deal of ore. Boyd set out to enter into a branching tunnel when his ears were suddenly assaulted by "a noise like the bursting of a thousand cannons ... followed by a most terrific rush of air which came from the drift that I was about to enter."

Terror in the Mines!

The rushing air lifted Boyd off his feet and threw him against the rock walls of the shaft with such force that he was nearly knocked senseless, and he was badly bruised. After the rush of air subsided, there came "one of the most piercing and plaintive cries I have ever heard ... a wail that produced indescribable and uncontrollable terror...."

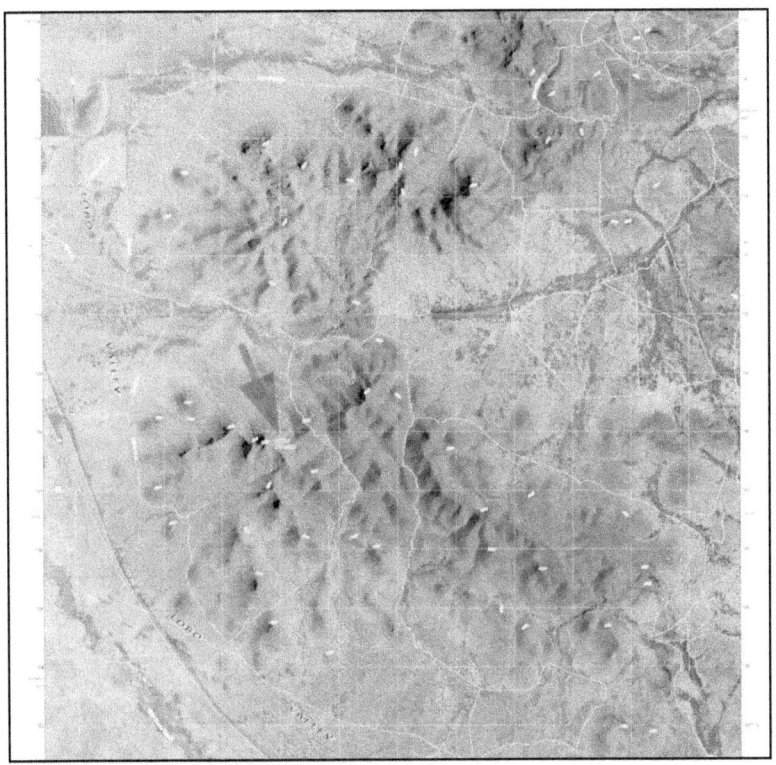

USGS Satellite Map of Chispa Mountain.

Terrified, Boyd got onto the rope ladder and screamed at his guide to pull the ladder up. However, the guide had heard the commotion in the shaft, causing him great fear, and had run off, leaving the rope ladder attached to a nearby tree trunk. With nobody to help him out, Boyd carefully climbed the rope ladder back to the surface.

Following this incident, word spread all around the area that the mine was haunted. Evidently, the theory was floated that the mine was possessed by the unseen spirits of the Spaniards who worked the mine in the 1700s.

A member of a later exploration party said, "The men claim that … the ghost of an old Spanish conquistador, clad in the armor of his time, sprang at them from out of the gloom, uttering a wild shriek of anger."

Nonetheless, Boyd remained eager to get a party together and go back for more exploring. But, before he could organize an expedition, he died.

James E. Meade, a close friend of Boyd, decided to carry on with the expedition at some point after Boyd's death. Meade took a group of men to the abandoned silver mine, and the men sank a shaft to a depth of 50 feet when they began experiencing the loud noises that Boyd had previously told Meade about. "The noises became so pronounced that the workmen refused to go on with it, and the whole project was abandoned."

This strange newspaper story, which first appeared in December 1902 and continued being reprinted through 1905, said, "Since then many attempts have been made to explore the mine, but the experience has proven more than any man is willing to stand a second time."

Another attempt to pierce the secrets of the mine was undertaken by Captain Louis Sefton, "one of the most prominent stockmen in Texas." Leaving his ranch in Sutton County with half a dozen of his best cowboys, Sefton headed to the mine. Since the account was included in the very first newspaper article (1902), one must assume that Sefton's expedition occurred sometime between 1882 and 1902.

Sefton and two of his men descended the rope ladder and were about to enter a horizontal passage when the phenomenon of the strong wind suddenly broke out in full fury. "The three men were hurled with great force several feet and thrown repeatedly against the jagged rocks of the shaft. It was only with the greatest effort that they could climb to the surface. Their bodies were covered with bruises, and their clothing was torn."

Sefton concluded his account to the newspaper by saying, "I am not superstitious, but if the interior of that mine is not an inferno occupied by hellish spirits, I won't believe that I see with my own eyes hereafter."

Interestingly, a short blurb in the April 14, 1905 edition of the *El Paso (Texas) Morning Herald*, argues against any superstitious explanations for the various incidents at the mysterious mine, saying that a group of El Paso mining men "expect to make a thorough exploration of the caves and air passages, which they suspect is the cause of the strong air currents occurring intermittently at the bottom of the shaft." The men were said to also "strongly suspicion that the terrible piercing scream, if any, comes from one or more of the big cat family which make their headquarters in such places and gets in and out through passages coming to the surface further down the mountain side."

Yet another theory for the phenomenon was posed by the editors of the *Dublin (Texas) Progress* in their January 23, 1903 edition, which said, "Whenever the attempt has been made to explore [the mine], the explorers have been met by a sudden and violent blast of wind, accompanied by dreadful howlings and wailing suggestive of lost souls. The mine probably taps a hidden reservoir of air controlled by an underground stream of water and belongs to the class of 'blowing caves' found in different parts of the world."

Although the *El Paso Morning Herald* article mentions that the El Paso mining team would visit the abandoned mining site in May 1905, spending an entire month exploring the area, no update to their expedition was ever published in any newspaper account that we were able to find. At this point, it seemed the story of the "haunted mine" had come to an end. Except that it hadn't.

On May 11, 1907, two years after the El Paso expedition was supposed to take place, a newspaper article appeared in the *Chicago (Illinois) Saturday Blade* with a dateline of Marfa, Texas, which is 25 miles west of Alpine. The same article was reprinted a month later by the *Bryan (Texas) Morning Eagle*. This story told of yet another expedition to the haunted mine, this time led by a man named Tom Borgus, who presumably was from Marfa. Whether this final recorded expedition was related to the planned 1905 El Paso venture is unknown.

In this latest installment of the legend, the name of the haunted mine, previously called "Refugio," meaning "refuge," has now changed to "Muerta," which means "dead woman." The location of the mine is now given as "60 miles southwest" of Marfa, which is significantly nowhere near the previous location of Chispas Mountain. Sixty miles southwest of Marfa is the Texas border town of Presidio, which has had virtually no mining activity in its history. But 40 miles southwest of Marfa is the famous silver mining town of Shafter, which would have been a more likely possible location for the mysterious Spanish mine.

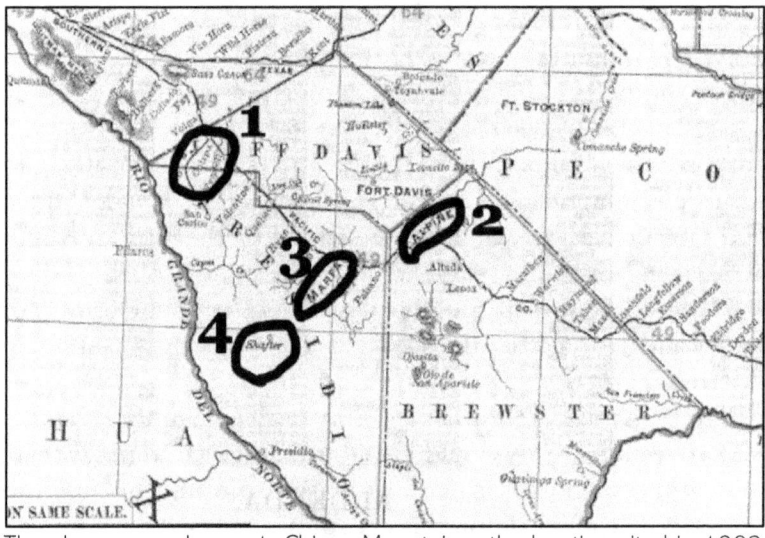

The above map shows: 1. Chispa Mountain – the location cited in 1902. 2. Alpine, Texas. 3. Marfa, Texas. 4. The vicinity of Shafter, about 40 miles SW of Marfa.

Like the previous story involving Henry Boyd, James E. Meade, and Louis Sefton, the Borgus story tells of an expedition to the haunted mine being organized. Borgus claims to have had great difficulty rounding up locals to be a part of his excursion. "The mine is the terror of the Mexicans of the upper portion of the Rio Grande border, and they cannot be induced to visit the locality, where it is situated, much less enter its dark and gruesome shaft."

Terror in the Mines!

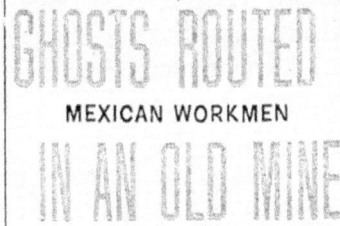

GHOSTS ROUTED
MEXICAN WORKMEN
IN AN OLD MINE

Owner Unable to Get Men Who Are Brave Enough to Work in Famous Haunted Mine.

MARFA, Texas, May 10.—Tom Borgus came in from the Chinati Mountains, sixty miles southwest of here, this week and laid in enough camp supplies to run him and his men several weeks. He is trying to reopen the famous Muerta mine, which is said to be haunted. The mine is the terror of the Mexicans of the upper portion of the Rio Grande border and they can not be induced to visit the locality where it is situated, much less enter its dark and grewsome shaft.

Borgus is an old mining man. He has prospected all over the West and Southwest and in many mining districts of Mexico. He was in Chihuahua a few months ago when the story of the Muerta mine was told to him by a Mexican who was down there from Presidio del Norte. Borgus made up his mind to take hold of the property. He came to Marfa and had no trouble in obtaining title to the mine. He spent some time trying to organize a working force among the Mexicans of this section. Not a man could be found who was willing to go with Borgus to the mine and help him work it.

Finally Borgus went over into New Mexico and employed twenty-five Mexicans to work in the mine. He said nothing to them about the shaft and underground workings being haunted. He brought these laborers to Marfa by train and rushed them out to the mine without giving them an opportunity to talk with any of the local Mexicans.

In less than a week the whole body of laborers were in Marfa again. They had deserted their work despite the vehement objections of Borgus. When questioned as to the cause of their quitting the job the Mexicans shrugged their shoulders and answered:

"Ghosts!"

The leader of the gang of Mexicans told a thrilling story of the adventures of the men. He said that the shaft is about 100 feet deep and at its bottom a shaft or tunnel extended to a chamber of con-

(Continued on Page Two.)

GHOSTS ROUTED WORKMEN
Owner Unable to Get Men to Work in Famous Haunted Mine.

(Continued from Page One.)

siderable size from which it appears a considerable quantity had been removed at some period, in the history of the property. This is evident from the fact that a large pile of rich silver ore was found at the far end of the chamber.

Soon after their arrival at the mine Borgus set the men to work cleaning out the shaft and straightening up the decaying timbers. Some of the Mexicans remarked while employed in this work that they heard strange rumblings which seemed to come from the interior of the mountain. Borgus laughed away their fears and told them the noises were only the echoes of the sounds from their own picks.

When the shaft was cleaned an exploration of the tunnel and chamber was made. Borgus set the men to work carrying the ore which was in the chamber to the surface. Shortly after they started in on this the skeleton of a man was uncovered from the rubbish. This grewsome sight unnerved the superstitious Mexicans and a few minutes later when a sound which seemed to come from under their very feet was heard and was accompanied by a strong gust of wind that extinguished the lights every man of them dropped his tools and fled in terror out of the chamber and up the ladder in the shaft to the open air above.

Borgus was at the surface and witnessed the exit of the men with no little dismay. He could not induce them to return to work and despite his entreaties they all left the place and came to Marfa. The men claim that when the skeleton was disturbed the ghost of an old Spanish conquesadore, clad in the armor of his time, sprang at them from out the gloom uttering a wild shriek of anger. Terrified, the workmen fled screaming to the opening of the mine.

The ancient mining records on file in the archives of the State of Coahuila, Mexico, at Saltillo, are said to show that the Muerta mine was operated by the Spaniards more than a century ago. It was abandoned on account of the mysterious and weird demonstrations that were said to have occurred in the underground workings.

Chicago (Illinois) Saturday Blade (May 11, 1907).

In this version of the story, Borgus found out about the haunted mine when he was visiting Chihuahua, Mexico, in 1906. He said a Mexican told him the legend, and he made up his mind to return to Texas and acquire the property, which he did.

Borgus finally succeeded in getting an expedition together, utilizing manpower that he recruited from New Mexico, where the haunted mine legend was not known. He found 25 Mexican men willing to work for him and rushed them to the site of the mine before they could talk to any of the local people and find out about the legend.

The leader of the workmen later stated that they had gone down the entrance for 100 feet and found "a large pile of silver ore." Borgus had them clean out the shaft and straighten up the decaying timbers. Afterward, they began carrying the ore up to the surface. Although they heard some disturbing sounds while working in the mine, including strange rumblings, Borgus persuaded them to continue.

At this point, while removing the ore, they uncovered from the rubble the skeleton of a man. "This gruesome sight unnerved the superstitious Mexicans, and a few minutes later, when a sound which seemed to come from under their very feet was heard and was accompanied by a strong gust of wind that extinguished the lights, every man of them dropped his tools and fled in terror...."

The men raced back to Marfa and spread the word that when they had disturbed the skeleton, there had appeared the ghost of an old Spanish conquistador, clad in the armor of the time, which sprang at them from out of the gloom, uttering a wild shriek of anger."

The article concludes by noting that the existence of the mine was first discovered in the archives of the State of Coahuila, Mexico, at Saltillo – just has had been reported in the original 1902 article. But it added a detail not previously told – that the Spaniards had abandoned the mine in the 1700s due to "mysterious and weird demonstrations that were said to have occurred in the underground workings."

Although this story seems strangely compelling, there are many elements that are very troubling about it. Searches of U.S.

Terror in the Mines!

Census records and genealogical records failed to turn up information on the persons named in these accounts – Henry Boyd, a railroad engineer residing in San Antonio in 1882; James E. Meade, also living in San Antonio at the time; Captain Louis Sefton, supposedly a prominent stockman who owned a ranch in Sutton County, Texas; and Tom Borgus, presumably a resident of Marfa. It seems unusual that we found no mention of any of these individuals in periodical searches for this time period, except in connection with this one story about the alleged haunted silver mine in West Texas.

Another disturbing fact is that the original newspaper article that was reprinted by so many papers from 1902 to 1905, including several Texas newspapers, seems to have originated with the *Saturday Globe* of Utica, New York. Why would an article about something that happened within 60 miles of Alpine, Texas, not be printed in the Alpine newspaper until a month after it was published in the Utica *Saturday Globe*?

Finally, there are errors of geography. The original location of the Chispa Mountain is given as sixty miles southwest of Alpine, when it is actually northwest. Also, although much mining has been carried out in the general region, there appears to be no historical records of any mining activity in the specific area of Chispa Mountain. Additionally, the 1907 article changes both the name and the location of the mine.

Is it possible that the prospectors involved in trying to exploit the mine were purposely trying to hide the exact location of the mine? We suppose so. Also, all of the locations mentioned are relatively close to each other – within perhaps an 80-mile radius. So, it is conceivable that the general geography mentioned in the articles could be within the realm of possibility.

However, due to all of these lingering questions, the authors are forced to hang a huge question mark in regard to the truthfulness of this story. The question mark must remain at least until such time as more information can be uncovered that corroborates the claims made therein. Sadly, that is the note upon which we must end this fascinating tale.

SMELTER IS HAUNTED

GHOST OR DEVIL APPEARED AT LANYON SMELTER SATUR- DAYNIGHT TWICE.

SCARED NIGHT MEN TO DEATH

The Thing Appeared at Furnace No. 9 Works No. 1—Had oHrns and Its Head Floated In Air—Man Who Saw It Collapsed In a Heap and Daylight Was Awaited Before Night Force Went Home.

A ghost appeared last Saturday night in furnace No. 9 at Works No. 1 of the Lanyon Zinc Company, and to this day nobody knows the truth about it.

The night shift was working at the furnace when the Thing appeared just about midnight. When first seen it was standing beside the furnace, between No. 9 and the adjoining one. The man who saw it gave a yell and fled. The other men ran to the place to see what was wrong and all took one look at the Thing and away they went. The front man fell down and the others fell over him. Afterward they screwed up their courage and returned and it was gone.

Little else was talked of for some time. The Thing had horns and long hair, great big eyes and an inhuman look, although standing erect like a man. It was a Spirit or the Devil. Before the excitement had subsided one of the men went to a window at the west side of the furnace room and opened the window. Merciful heavens! There it was! The man sank in a heap. Another workman walked over and threw a chistle at the Thing's head. It did not dodge, nor run, but came forward. When it disappeared again it stayed away. The night metal drawers and stuffers, although through their work at 3 o'clock, remained until daylight before going home.

Among the men working that night were Chris Klamick, who was overcome; Ed. Mundis, John Rice and Len Hartley were also on the shift. The boys have kept very quiet about the Thing, but they won't deny that it scared the livers out of them, and they have a constant fear that it may come back.

Whether the apparition is, the Evil Spirit of the Gas, or some man masquerading for the fun of it, the men neither know nor care. But they know the Thing—the inhuman, awful Think—stood there. Ghost, Devil or man, they do not care to renew acquaintance.

Iola (Kansas) Daily Register and Evening News, Nov. 18, 1903, p. 1

3.

THE THING IN THE MINE

NOVEMBER 14, 1903
IOLA, KANSAS

To this day, nobody knows what sort of creature had made its home deep in the zinc mines of Iola, Kansas, in November of 1903. The horrified miners described it as humanoid in appearance and somewhat devil like! The bizarre apparition was first spotted by the zinc miners in the vicinity of Furnace Number 9 of the Lanyon Zinc Smelter on November 14, 1903.

The newspaper account in the November 18 edition of the *Iola (Kansas) Daily Register and Evening News* stated that the night shift of miners was hard at work at the Lanyon smelting facilities when something unbelievable happened. It was near midnight on a Saturday, when several of the workers noticed a tall humanoid standing between two of the furnaces used to smelt zinc ore.

Referred to simply as "The Thing," the strange creature stood erect like a man, sported long hair, looked at the miners through huge eyes, and had horns atop its head! The men who first saw the creature let out a chorus of yells, which brought the other smelter workers running to see what the commotion was about.

Once all the men got a careful look at what they were dealing with, a mad scramble to exit the smelter ensued. One unfortunate miner stumbled, falling upon the ground, and was mercilessly trampled by his colleagues as they speedily vacated the premises.

For quite some time after the encounter, the workers talked about nothing else, discussing at length what they had witnessed. The newspaper account said, "Little else was talked of for some time. The Thing had horns and long hair, great big eyes and an inhuman body although standing erect like a man. It was a spirit or the Devil."

Zinc smelter workers in Iola, Kansas – Date Unknown
(Courtesy EPA.gov)

Motivated to solve the mystery, the miners "screwed up their courage" and returned to the area of their earlier sighting. But the terrifying creature was gone.

As the evening wore on, one of the workers, Chris Klemick, was opening one of windows of the furnace room, when he saw the creature once again. Klemick immediately fainted. One of the other workers picked up a heavy chisel and heaved it at the Thing's head. "Before the excitement had subsided, one of the men went to a window on the west side of the furnace room and opened the window. Merciful heavens! There it was! The man sank in a heap. Another workman walked over and threw a chisel at the Thing's head. It did not dodge, nor run, but came forwards. When it disappeared again, it stayed away," according to the newspaper account.

Zinc smelting operation in Iola, Kansas – Date Unknown (EPA.gov)

The newspaper noted that even though some of the smelter workers ended their shift at 3 a.m., they did not leave the zinc works for fear of running into the creature on their way out. Instead, the workers stayed indoors until daybreak before finding the courage to return to their homes.

The miners involved in the encounter were actually identified by name in the newspaper article: "Among the men working that night were Chris Klemick, who was overcome; Ed. Mundis, John Rice and Len Hartley were also on the shift."

For a time, the workers kept quiet about their experience. "The boys have kept very quiet about the Thing, but they won't deny that it scared the livers out of them, and they have a constant fear that it may come back."

Added the paper, "Whether the apparition is the Evil Spirit of the Gas, or some man masquerading for the fun of it, the men neither know nor care. But they know the Thing – the inhuman, awful Thing – stood there. Ghost or Devil or man, they do not care to renew acquaintance." The newspaper article concluded, "To this day, nobody knows the truth about it."

No. 1 SMELTER, LANYON ZINC CO., IOLA, KAS.

Photo of the Lanyon Zinc Works, Date Unknown.

What was this strange creature? Was it a denizen of outer space or the inner Earth? We know of several interesting UFO cases in Kansas during the 19th century – so perhaps the creature was left behind or missed its ride back home? A search of the historical record for more sightings of this strange humanoid resulted in no further data.

4.

ALASKAN MINER SEES FLOATING CITY

JUNE 1888
MUIR GLACIER, ALASKA

Toward the end of the 19th century, Richard G. Willoughby, the first American prospector to find gold in the Klondike, was working in the wilderness at Muir Glacier, Alaska, when he claimed to have seen an incredible sight floating in the sky above the glacier. The image was that of a wonderous city, which supposedly was later found to be visible on a regular basis in the same area.

A later eyewitness to the phenomenon said, "Whether this city exists in some unknown world on the other side of the North Pole, or not, it is a fact that this wonderful mirage occurs from time to time yearly, and we were not the only ones who witnessed the spectacle. Therefore, it is evident that it must be the reflection of some place built by the hand of man."

The story of Willoughby's "mirage city" even appeared in the June 1897 edition of *Popular Science Monthly*, which said, "Mr. Richard G. Willoughby is a mining prospector and 'promoter,' resident in Juneau, Alaska, a man whose vocation enables him to see some wonderful things. In June, 1888, according to his statement, Mr. Willoughby beheld an extraordinary mirage from the surface of the Muir Glacier. It was the apparition of a great city of tall houses of brick and stone, plainly shown in the air under the influence of some powerful refraction. Behind the city was a river in which shipping was faintly shown. In the foreground the leafless branches of tall elm trees were clearly traceable. In the center of the city was a large edifice with several towers, and on some of these towers the presence of scaffolding showed that building was still going on. This mirage was seen by him several times from year to year, and on the unfinished building the stages in the process of erection each season could be distinctly followed."

Muir Glacier, photographed in 1897.

The strange apparition is also mentioned in Alexander Badlam's 1890 book *Wonders of Alaska*, (1890, p. 127-131), in which Badlam describes "Willoughby's Mirage" on Muir Glacier. Also, the sighting drew the attention of the American paranormal researcher Charles Fort, who included an account of it in his 1923 book *New Lands* (pages 491 and 492). Fort quoted from an article in the *New York Tribune*, which "told of an occasional appearance, as if of a city, suspended in the sky, and that a prospector, named Richard G. Willoughby, having heard the stories, had investigated, in the year 1887, and had seen the spectacle. It is said that, having several times attempted to photograph the scene, Willoughby did finally at least show an alleged photograph of an aerial city.

Photograph supposedly taken by Willoughby.

The general feeling was that somehow they were witnessing the image of a city located hundreds or thousands of miles away, being somehow mysteriously projected onto the sky above them. "Not one of us could form the remotest idea in what part of the world this settlement could be. Some guessed Toronto, others Montreal, and one of us even suggested Peking."

However, observers noted that the image in the sky did not look like any of the places mentioned as the possible sources. They agreed that it looked like a manmade city, rather than something out of this world. One witness said he thought it looked like "some immense city of the past" – almost as if they were looking through a time anomaly that gave them a view of years gone by.

Journalist Minor W. Bruce said that Willoughby had told him of the phenomenon, and that, early in 1899, he had accompanied Willoughby to the place over which the mirage was said to repeat. It seems that he saw nothing himself, but he quoted a member of the Duc d'Abruzzi's expedition to Mt. St. Elias, summer of 1897, Mr. C. W. Thornton, of Seattle, who saw the spectacle, and wrote – "It required no effort of the imagination to liken it to a city, but was so distinct that it required, instead, faith to believe that it was not in reality a city."

Unfortunately for Willouhby and the other believers in the fabled city in the sky, *Popular Science Monthly* had an expert look closely at the photograph that Willouhby claimed to have taken of the city and concluded that it was a fraud.

Richard Willoughby (Alaska State Library).

In its June 1897 edition, *Popular Science* disclosed: "Prof. William H. Hudson, of Stanford University, who lived for a time in Bristol, England, recognizes the picture as a view of that city from Brandon Hill, above the town. The picture must have been taken some twenty years ago, because Prof. Hudson distinctly remembers the scaffolding around the towers of Bristol Cathedral at that time while the building was being repaired. The hotel and the church to the left of the cathedral are also recognized by him."

Terror in the Mines!

The conclusion of Hudson's analysis was that the supposed photograph of the magical floating city was a fake! "A more transparent fraud could hardly be devised, but its very imbecility assures its success. We may be certain that for many years to come the 'Silent City' will be the 'wonder and pride of Alaska's bleak hills,' and tourists eager to 'pierce the veil' will speculate on the probability of its being 'perhaps altogether within the recesses of another world,'" said the magazine.

While the photograph seems to have clearly been faked, could it be that Willoughby and possibly others actually saw something unexplained in the sky above the Muir Glacier? In a 2018 article by Alex McCarthy of the *Juneau Empire*, he recounts the entire episode of Willoughby's "mirage city" and concludes by saying, "It's not clear whether Willoughby ever admitted to the hoax."

Mining Operation in Creede, Colorado (Library of Congress).

5.

VANISHING FIGURE IN COLORADO MINE

1903
CREEDE, COLORADO

In early 1903, a mining engineer named Neil McQuig (misspelled as "McQueg") told the *Denver Times* of a ghostly incident that happened to him while working at a mine in Creede, Colorado. McQuig said that on June 24, 1902, he was working inside a mine shaft on the property of the Big Kanawha Mining Company at Creede. While standing at the end of the tram, he noticed a man standing "not more than 20 feet away." Thinking the man was a miner, McQuig called out to him twice but received no reply.

Upon moving toward the figure, the apparition immediately vanished "as if by magic," and not the slightest trace of a man could be found. None of the men working in the area had seen anything of the stranger, although they were not close to where McQuig saw the figure.

Moments later, the men working in the mine shaft, including miner Henry S. Jones, were startled to hear the three distinct signals instructing them to evacuate the mine. It was discovered later that nobody had given the signals and that there were no workers within 15 feet of the mechanism for giving the signal.

Three days after this unexplained event, Henry S. Jones slipped and fell 800 feet to his death, down the same shaft from which the mysterious signals had come.

Mining Operation in Creede, Colorado. (Library of Congress)

Following this incident, miners began telling stories of ghostlike figures seen at the entrance of the mine shafts. They told of hearing groans and moaning sounds coming from the bottom of the shafts. "These tales ... are said to have become so impressed upon many of the men employed in the Big Kanawha company's mines that they have quit work and sought places in other mines, where the unnatural sights and sounds are unknown."

Although the nature of the strange apparition seen by McQuig cannot be known for certain, the authors have been able to verify that the persons mentioned in this story were real people. For example, a death notice was found for Henry S. Jones, described as a "pioneer miner of Creede camp whose wife and baby reside in Denver." The notice stated that Jones fell from the top of the "Happy Thought skip" to the eighth level, a distance of 800 feet. The fall killed him instantly, mangling his head beyond recognition. (In mining, a "skip" is a mechanical conveyance used to transport mining material to the surface.)

HAUNTED MINE

Ghostly Forms Scare Workers and Gives Mysterious Signals.

Weird stories of ghostlike figures seen at the entrance of the mine shafts and tales of groans and moaning sounds being heard from the bottom of the shafts are related by mining men who have just returned from the property of the Big Kanawha Mining company at Creede. These tales, which rival those of "Babletts," of whom Frank Daniels sang, are said to have become so impressed upon many of the men employed in the Big Kanawha company's mines that they have quit work and sought places in other mines, where the unnatural sights and sounds are unknown.

The first man to relate a ghostlike tale of his experiences while working at the mine was Neil McQueg, an engineer. It was about five months ago that while stranding at the end of the tram he saw a man not more than twenty feet away. Thinking it was one of the miners employed at the place, he spoke to him. He received no reply, and again he addressed him. Again he received no reply.

Terre Haute (Indiana) Daily Tribune, 2-11-1903, Page 3.

and this time McQueg determined to find out who the man was. He approached to where the figure had been, and as he neared the spot the man disappeared. No trace of the man could be found. McQueg swears that he saw a man, but the moment that he approached to where the figure had been it vanished as if by magic. Not even the slightest trace of a man could be found, and none of the men working around the mine at the time say anything of a stranger, nor were any of them near the place where McQueg had seen the figure at about the time that the engineer saw the stranger.

Shortly after this three distinct signals given to hoist the men from the station were heard in the bottom of a shaft where Henry S. Jones and some others were working. At the time the signals were given no one was nearer the place than fifteen feet, and the search made to discover the person who gave the signals has been unavailing.

Three days after this strange occurrence on June 26, 1902, Henry Jones was killed by falling out of the bottom of the skip a distance of several hundred feet down the same shaft from which the mysterious signals to hoist the men had been given.

These three occurrences made a great impression upon many of the men and particularly the more superstitious of them.

Also confirmed to have been an actual person is Neil McQuig, born in Canada in 1858, and residing in Ouray, Colorado, at the time of the 1900 U.S. Census. He is listed as a single man living in a boarding house. During the time of this incident (1902), McQuig would have been 44 years old. He is listed in the census as a "stationary engineer," which is a person who operates and maintains boilers, engines, air compressors, generators, and other equipment. The term "stationary" comes from the fact that the engineer usually remains fixed in one location, from where he can monitor the equipment.

> At Creede on June 27th Henry S. Jones, a pioneer miner of Creede camp, whose wife and baby reside in Denver, fell from the top of the Happy Thought skip to the eighth level, a distance of 800 feet. The fall killed him instantly, and mangled and bruised his head beyond recognition. Jones fell through the bottom of a self-dumping skip.

Meeker Herald (July 5, 1902, page 3).

What happened to McQuig after his specter encounter in June 1902 is unknown. No further mention of him was found in the historical record, and it is possible that he returned to his native Canada at some point after 1902. A search for death information about McQuig was unsuccessful.

6.

HELLHOUNDS OF HELLDORADO

1850S-PRESENT
ELDORADO CANYON, NEVADA

In literature and mythology, a hellhound is a large, fierce dog-like creature that guards the entrance to hell or the underworld. The legend of these spectral hellhounds occurs in almost every country and geographical region on Earth. Although descriptions of these creatures vary, they are generally black, tremendously overgrown, supernaturally powerful, and often have red eyes or are accompanied by flames. One of the greatest of all Sherlock Holmes stories, *The Hound of the Baskervilles*, concerns the appearance of a creature that is believed to be a hellhound.

It is not surprising that the European legends of these fierce, supernatural dogs made their way to America and surfaced among the miners of the Old West. Sometimes a certain mine was said to be haunted by one of these spectral beasts. Superstitious miners would catch glimpses of a canine shape moving in the darkness. When miners were found dead of unexplained causes, survivors would float the theory that they had been killed by a hellhound.

The best-known hellhound tales of the Old West arose around the mining settlements of Eldorado Canyon, Nevada. The area's name, meaning "Gold Canyon," is ironic because Spanish explorers named it long before any gold was ever discovered there. The Spaniards found only veins of silver, but in the 1850s, a group of prospectors finally did discover gold there, and by 1858, as more people made their way up to the area, gold fever broke out with full force. Very quickly, El Dorado became one of the wildest spots of the Wild West.

Gold Miners in Eldorado Canyon, c.1850.

It was in this lawless frontier area, where murders happened daily and lawmen looked the other way, that stories of paranormal happenings, including hellhounds, sprang up. According to Kathy Weiser in *Legends of America,* "Man per man and mile per mile, Eldorado Canyon has a wider range of historical events than anywhere in the Wild West. This rich history, coupled with the turbulent events taking place in Eldorado Canyon in the 19th century has led to numerous ghost stories of dead miners, Indians, and pioneers who once roamed the area."

Among the most famous of Eldorado's ghosts are easily the hellhounds. Miners caught fleeting glimpses of large dog-like beasts darting among the boulders. Shadowy figures seen at night caused terror among the settlers.

Skeptics believe that residents of Eldorado were actually seeing wild dogs that had previously been used by miners to help protect their claims and were later abandoned to run wild. It is a historical fact that many prospectors in the area kept vicious dogs chained up at their camps to keep claim jumpers away. These dogs were typically quite antisocial, much like the junkyard dogs of later generations. Sometimes they escaped or were abandoned by the miners when their work at the site was completed.

Today, paranormal investigators report that spectral black dogs are the most commonly seen "ghosts" in Eldorado. A user on *Shadowlands.net* wrote of seeing the legendary hellhounds in the area. The author and his brother had heard of the legend and decided to explore the Eldorado region north of Hoover Dam. Their first few trips didn't yield any evidence of the dogs, spectral or otherwise. On their final visit, they found an eight-foot chain embedded into the rocks at the entrance to a mine shaft. The brothers entered the mine shaft and found the bones of a very large dog. But still, no specters showed up. Either feeling confident that there were no ghosts, or perhaps hoping still to see one, the brothers made camp outside of the mine shaft. That night, they could hear the howls of what they assumed were coyotes when suddenly the atmosphere became thick and very uneasy.

"We now felt that we were being watched from a very close distance. What we thought was the night time breeze now sounded more like the panting or breathing of large dogs in close proximity. Then we heard the growling. Grating, low......and hateful. The fall of paws on the desert sand now became apparent. They seemed to circle the campsite. We were surrounded."

From a boat in midstream looking west.

Southwestern Mining Co. Quartz Mill.   Mouth of El Dorado Canyon   Boarding/Store/Niffemen's Weather House /cabins /Observation Station

Eldorado Canyon from the Colorado River c.1900-1925.
(UNLV Special Collections)

Next, the brothers' attention was drawn to a scratching noise coming from the entrance to the mine. They looked and could see the chain moving as though an invisible dog were attached to it. The chain began to tug away from the rock it was attached to while at the same time, scratch marks and blood began to appear on the rock.

Finally, a hairy invisible something brushed against the author's leg. The brothers had gotten the ghostly encounter they had hoped for and promptly ran for their car. All the while, they could hear the ravenous panting and the rhythm of dogs' feet pounding the dirt as they ran to the vehicle. The two men made it to their car and sped away. It was at this point that the ghostly dogs either materialized to become visible, or a live pack of dogs began following the car:

"On the road heading out of the canyon we were paced for a good two or three miles at least by what seemed to be a pack of wild strays! We made it home and I will never forget the terror of being chased by this pack of spectral hounds...NEVER!"

While an interesting story, it is actually only one of a few as it turns out. Truthfully, not as many people claim to see these "hellhounds" of Eldorado as you might think. Or, if they do, they mostly go unreported. Apart from the account published on *Shadowlands*, only one other hellhound account can be dug up that relates to Eldorado. The ever-dependable Brent Swancer of *Mysterious Universe* found an account on the now-removed *King Sasquatch Paranormal & Cryptozoology Blog*.

The tale told of a group of friends out four-wheeling in Eldorado Canyon. One member of the group saw what he took to be a coyote crouching in a "defensive stance" while his friends next to it could see no such thing (the implication being that he was seeing a ghost dog while they were not). Later that night, the shadow of a canine crept across the tent of one of the girls there, who screamed loudly at the sight of it. It is said that when she did, the animal disappeared, but the poster wasn't specific enough in their wording to clarify if he meant it ran away or faded away like a specter.

The same blog had one other account in which the witnesses were boating up the Colorado River in the vicinity of Eldorado Canyon. This account was a bit more interesting as it implied more of a flesh and blood cryptid rather than a ghost. The account was related by the father of one of the witnesses, who wrote:

"Around two in the morning my house phone rang off the hook with my son fanatically shouting that he had just seen a mutant dog with a piercing howl attempting to catch a duck. He forwarded the details as a four-foot mangy dog with terrifying overlapping teeth. He said the dog failed to catch ducks and ran off hungry when they shined a flashlight onto the shore."

If not for its having been seen in the haunts of the hellhounds, one might well have lumped this sighting in with that of the mangy Chupacabra. Was it perhaps one of the living descendants of the prospectors' guard dogs set free into the wilds of the canyon, or did one of the spectral dogs manage to make a physical manifestation of itself?

Though sightings of these hellhounds are relatively few in number, due to their unique history, their legend has caught on quite well. They even attracted the attention of Jack Osbourne for his series *Haunted Highway* in 2102. In the episode where Jack and co. visit Eldorado Canyon, their thermal imaging cameras appeared to pick up a dog-like quadruped in the darkness, approaching one of the show's cohosts. In true *Blair Witch Project* style, as the cohost runs through the desert, for a split-second, a dog-like beast is seen to jump across the screen and hit her on the shoulder and then disappear. Was it simple TV trickery or a real hellhound sighting? Considering that specters and cryptids are loathe to show up for those actually hunting them, the former seems more likely.

7.

THE MONSTER OF THE JORNADA DEL MUERTO

JANUARY 1888
JORNADA DEL MUERTO, NEW MEXICO

Imagine traveling along one of the most rugged and desolate stretches of desert in New Mexico when you are suddenly confronted by a sixty-foot-long dragon-like beast. This story was reported in the January 27, 1888 edition of the *Helena Independent* newspaper, based on a "dispatch" received from San Marcial, New Mexico, about 20 miles south of Socorro. The creature encounter is said to have happened near an extinct crater on the plain known as the "Jornada del Muerto."

Jornada del Muerto, translated as "Journey of the Dead," remains even today almost entirely uninhabited and undeveloped. This desolation stretches from north of Las Cruces to south of Socorro, New Mexico. So isolated is this area that it was chosen as the "Trinity Site," where the first detonation of an atomic bomb took place in 1945. In recent years, it is the site of Spaceport America, an FAA-licensed spaceport located on 18,000 acres, from where a Virgin Galactic spacecraft carrying three persons aboard was first launched in 2021. It seems fitting that this mysterious and desolate area should be the location of one of the strangest "unexplained creature" encounters of the 1800s! As the story goes, back in January 1888, the first sighting of the beast was reported by a landowner identified only as "Mr. Alexander," who owned several mining properties in the nearby San Andres Mountains.

Alexander was crossing the Jornada on his way to inspect his mines when he had his encounter. He was on foot, jogging behind his burro, when suddenly the burro stopped, raising its long ears as if listening to something. Then, the burro quickly turned around and took off in "a mad stampede."

Jornada del Muerto. (U.S. National Park Service)

Alexander soon saw what had spooked the animal, and his nerves became "completely paralyzed, his hairs stood on end, and move he could not." About a quarter of a mile from the spot where Alexander stood was the strangest apparition he had ever seen – sort of a cross between a dragon and a huge serpent!

He described it as about sixty feet in length with a head as large as a "barrel." The newspaper account said, "A few feet behind the creature's head two large scales were visible, which glittered in the sun like polished shields; further back were two huge claws on either side, about two feet apart, which were all the monster had in the shape of feet. The rest of its body was comparatively small and tapering to the end of its tail."

The strange creature, although serpent-like, travelled at a rapid pace, sometimes rearing up its whole body off the ground and walking on its four claws. The bizarre beast, which was travelling away from him at a fast clip, soon went over a hill and was lost to his view.

Terror in the Mines!

The newspaper reporter interviewed a number of area residents, described as "Mexicans," who said they were aware of a strange beast that dwelled in the place where Alexander had his encounter. The reporter said, "The Mexicans have the most deadly fear of the crater, and will not venture within miles of it, there being a popular tradition among them that it is the abode of some terrible serpent. The Mexicans asserted that on one occasion a descent of the crater was made by three men, and as none of them returned it was generally believed that they were devoured by the monster."

Artist's conception of mythological dragon.

The volcanic crater mentioned in the story was most likely Kilbourne Hole. Specifically, it is located between the Potrillo Mountains and the Rio Grande in southern Doña Ana County. Technically Kilbourne Hole is a maar, a depression or pit created by a volcanic explosion with little material emitted aside from volcanic gas. The maar is believed to be somewhere between 24,000 to 100,000 years old. The crater is elliptical in shape, 1.7 miles long, and hundreds of feet deep.

Considering that the newspaper story mentions a Mr. Alexander on his way to his mine, it's worth noting that the site of Kilbourne Hole was rich in unique minerals. Namely, the crater contained debris that when cracked open comprised sparkling yellow and green interiors of olivine glass granules. If Mr. Alexander mined these materials is unknown. In any case, in 1975, Kilbourne Hole was designated as a National Natural Landmark by the National Park Service.

MEXICO'S PRIZE SNAKE

It is a Monster, indeed, and it Abides in an Extinct Crater.

San Marcial Despatch: This section of the country has been considerably aroused from time to time by the conflicting reports of Mexicans, who say that the extinct crater to the east of the plain known as the "Jornado del Muerto," about twenty-five miles from this place, is the abode of a monster serpent, second in size only to that huge reptile of the seas that has so often been spoken of by mariners and others. It is reported by some to be fully 100 feet in length and about two feet in circumference, but probably the most trustworthy information is that given by a Mr. Alexander, who possesses some mining property in the San Andreas mountains, which he east of the broad plain. Mr. Alexander says that he saw the serpent once while crossing the Jornado on the way to his mines. He was about half way across the plain, jogging leisurely along behind his burro, dreaming of the immense wealth that he hoped to realize from his property, when suddenly the burro stopped, erected his long ears, wheeled quickly around and made a made stampede in the opposite direction. Mr. Alexander was at a loss to account for this strange freak of the burro, and was about to start in pursuit of the runaway, when he chanced to look ahead. Then his eyes gazed upon the monster. He was so beside himself with fear at first, he says, that his nerves were completely paralyzed; his hair stood on end and move he could not; he was rooted to the spot, and his eyes were fixed upon the serpent. It was about a quarter of a mile from him, and was traveling in the opposite direction—towards the crater. He says it appeared to be about sixty feet in length; but what surprised him most was the queer proportions of the creature. The fore parts were of enormous size, its head being fully as large as a barrel. A few feet behind the creature's head two large scales were visible which glittered in the sun like polished shields; further back were two huge claws on either side, about two feet apart, which were all the monster had in the shape of feet. The rest of the body was comparatively small and tapering to the end of the tail. It traveled at a rapid gait, sometimes rearing its whole body from the ground, and walked on its four claws. He watched it till it disappeared over a little hill, and then he started to look after his burro.

The Mexicans have the most deadly fear of the crater, and will not venture within miles of it, there being a popular tradition among them that it is the abode of some terrible serpent. The Mexicans assert that on one occasion a descent of the crater was made by three men, and as none of them returned it was generally believed they were devoured by the monster.

Helena Independent, January 27, 1888, p. 3

The crater mentioned could have also been Aden Crater, a small shield volcano (a type of volcano usually composed almost entirely of fluid lava flow). Like Kilbourne Hole, it too is a part of the Potrillo volcanic field of New Mexico. The crater is about the size of the Sun Bowl Stadium at the University of Texas at El Paso. Adding to the flavor of the story is the fact that the area has deep caverns that are hazardous to hikers. Furthermore, in one of the caves, a prehistoric ground sloth was exhumed.

So in the end, how likely is this creature sighting to be true? Since over a century has passed, it is impossible to follow up on many of the details mentioned in the newspaper account. Interestingly, the only newspaper account we could find appeared in a newspaper from Helena, Montana, which is over 1,000 miles away from where it happened. The Helena paper cited a dispatch from San Marcial, New Mexico, as the source of the story. San Marcial, now a ghost town, is about 25 miles south of Socorro. San Marcial's population in 1888 would have been about 1,000.

Although San Marcial had newspapers during this time period, news coverage was sparse and spotty. In looking through what few newspaper records remain, we found no mention of the dragon-like creature or the mysterious landowner named "Mr. Alexander."

The story itself seem plausible in that it mentions correct geographic details. So, although its veracity remains in question, we have no evidence at this time to conclude that the story is a hoax.

CURIOUS THRONGS FLOCK TO HAUNTED TREASURE SPOT

Stories of Eerie Wails, Invisible Hailstones and Dim Ghost Lights Come From Remote Corner of Oklahoma.

OKMULGEE (Æ)—That exclusive set of ghosts on Snake creek is going to have its hands full if it keeps away the army of curious who are flocking to the haunted treasure spot where they stand guard.

Stories of invisible put painful hailstones, eerie wails and dim ghost lights, that spectres of five Indians and one white man used to frighten away seekers after legendary gold buried in their graves on Snake creek's bank have aroused Okmulgee county.

The little farming community in the northeast corner of the county wears a pitying smile for those who do not believe and several reputable farmers of the neighborhood stand ready to convince the incredulous of the existence of the weird manifestations. One visitor to the ghostly spot is suffering nervous prostration.

Digging for Gold.

Digging for the gold has been going on in the day time, but those who touch pick or earth at night are said to hear blood curdling whoops from the five Indians.

The frightening screams are repeated from every angle and, if the visitor is not dislodged, lights are seen hanging from tree tops, strewn along the ground and even under the mortal's very feet, treasure seekers aver.

The climax is worth staying to see, Snake creekers says. A soughing gust of wind overhead grows into a rumble of thunder and small objects begin patting the adventurer from above. They feel like hailstones, but when they are caught, the hand is opened to find nothing. The ghostly bag of tricks may be deeper but none has stayed around to wait for the curtain.

Science has not yet undertaken to explain the phenomena.

The Idaho Statesman (February 14, 1927).

8.
SPECTERS OF SNAKE CREEK

FEBRUARY 1927
SNAKE CREEK, OKLAHOMA

Snake Creek is one of many small, rural settlements in Oklahoma. It began to be settled about the year 1910 via settlers from Texas. Though it was large enough to boast two different churches, there was never much to the settlement, and a devastating flood in the early 1940s incited quite a few residents to relocate. Snake Creek experienced its fifteen minutes of fame in 1927 via a lost treasure tale of dubious veracity. Apart from the splash made in 1920s era papers, the story of the haunted Snake Creek Treasure didn't seem to go too far after that.

"That exclusive set of ghosts of Snake Creek is going to have its hands full if it keeps away the army of curious who are flocking to the haunted treasure spot where they stand guard," began the *Casper Star Tribune* of February 13. The article proceeded to describe accounts of "invisible but painful hailstones, eerie wails and dim ghost lights, that spectres of five Indians and one white man used to frighten away seekers after legendary gold buried in their graves on Snake Creek's bank have aroused Okmulgee county."

Residents offered a "pitying smile" to unbelievers, and "several reputable farmers of the neighborhood stand ready to convince the incredulous of the existence of the weird manifestations." The paper even claimed that "One visitor to the ghostly spot is suffering from nervous prostration."

Apparently, the gold could be searched for in the daytime with no problems, but those who dared to dig at night heard "blood curdling whoops from the five Indians."

SNAKE CREEK GHOSTS ON GUARD OVER HAUNTED TREASURE SPOT

By Associated Press.

Okmulgee, Ok., Feb. 12.—That exclusive set of ghosts on Snake Creek is going to have its hands full if it keeps away the army of curious who are flocking to the haunted treasure spot where they stand guard.

Stories of invisible but painful hailstones, eerie wails and dim ghost lights, that spectres of five Indians and one white man use to frighten away seekers after legendary gold buried in their grave on Snake Creek's banks, have aroused Okmulgee.

The little farming community in the northeast corner of the county wears a pitying smile for those who do not believe and several reputable farmers of the neighbourhood stand ready to convince the incredulous of the existence of the weird manifestations. One visitor to the ghostly spot is suffering from nervous prostration.

Digging for the gold has been going on in the daytime, but those who touch pick to earth at night are said to hear blood-curdling whoops from the five Indians.

The frightening screams are repeated from every angle and if the visitor is not dislodged, lights are seen hanging from treetops, strewn along the ground and even under the mortal's verily feet, treasure seekers aver.

The climax is worth staying to see, Snake Creek says. A soughing gust of wind overhead grows into a rumble of thunder and small objects begin pelting adventurers from above. They feel like hailstones, but when they are caught the hand is opened to find nothing. The ghostly bag of tricks may be deeper but none has stayed around to wait for the curtain.

Science has not yet undertaken to explain the phenomena.

Galveston Daily News (February 13, 1927).

"The frightening screams are repeated from every angle and, if the visitor is not dislodged, lights are seen hanging from tree tops, strewn along the ground and even under the mortal's very feet, treasure seekers aver," the account continued.

The final act of the ghostly phenomena comprised a "soughing gust of wind overhead" that grew "into a rumble of thunder," after which an invisible hailstorm occurred. "They feel like hailstones, but when they are caught, the hand is opened to find nothing. The ghostly bag of tricks may be deeper but none has stayed around to wait for the curtain."

"Science has not yet undertaken to explain the phenomena," the article concluded.

When Wailing Cries Fail To Rout Greedy Treasure Seekers, Ghosts Pepper Them With Invisible Hail

Snake Creek, Okla., Citizens Decide to Hunt for Buried Gold During Daylight, Although the Spirits of Departed Warriors Place Eerie Lights on All Trees for Them.

Morning Examiner (February 13, 1927).

That brief article was the one that made the rounds in papers nationwide, a more detailed account of the story appeared in local papers, though, naturally. They also shed light on the ghostly apparitions.

The treasure was said to reside in a wooded area a little over two miles south of the county line near the Snake Creek school on the farm of Arthur Straughan. Oddly it was guarded by five Native Americans and one Anglo man, comprising six specters in all. Tales of the treasure spread to residents of nearby towns who came to investigate and were frightened away themselves by the ghosts. It also drew some spiritualists and ghost hunters, part of the spiritualism movement of the roaring twenties. One of the mediums declared that the six spirits were "sad, sad boys" who wished to be left alone. "In defense of their home the Indian ghosts have offered their best in screaming war whoops, while their white comrade flits about the tree tops lighting dim ghost lights," the *Morning Examiner* of February 13, 1927, reported. The paper also identified four witnesses in the form of Snake Creek farmers Hayworth Baker, who first found the haunted spot, and James Lee, Bob Benson and Earl Douglas. If the men were serious, or just in on the joke, is debatable. "Those men are well known and the stories they tell hang together well," as the paper put it.

"Speerits"

Jawn Barleycorn Upsets Snake Creek "Ghosts" With Theory of His Own About Visitations

A gentleman giving the name of John Barleycorn explains the Snake Creek ghost stories in a few well chosen words in a letter received yesterday.

There is no denial that there are spirits about Snake Creek, and the honor of a large number of men, who claim to have been terrified by manifestations of ghosts on guard over buried gold, is unsullied by Mr. Barleycorn's explanation. The beans are strewn in wild disarray.

John Barleycorn says:

"Sure their is spirits on Snake Creek plenty. If those ghost seekers are really hunting spirits let them take the price along and they sure will get them. A friend of mine got some last week for $7 a gallon those moonshiners dont want to many around at one time and a little of their own product in them will make them give a couple of war hoops sure it will those balls of fire they seen was the breath of the ghosts exploding."

Mr. Barleycorn certainly threw cold water on the party. And everybody was having such a good time.

This humorously slanted article from the *Okmulgee Daily Times* of February 17, 1927, asserted the whole incident stemmed from too much alcohol consumption.

The ghostly happenings began a few weeks previous to the articles, placing it in late January. The four men had heard tales of a treasure buried in the woods north of the school and set out to dig it up. The paper reported,

> The party had dug a few feet when a frightful Indian war whoop brought them to attention. It was not repeated, and they began again to dig. The screaming challenge was heard again, this time from several different directions. The men dropped their tools and fled in horror.

The men returned a few nights later when they had regained enough nerve to do so. At the instant the first shovel struck the ground, they heard another "Indian yell" which came from the treetops and from within the hole they were digging. It was at this point the men decided to consult with a spiritualist, or psychic, who went unnamed in the article. The psychic then claimed that it was the ghosts of five Indians and a white man causing the ruckus. The bodies of the six men had all been buried there years before and "rather liked the place and were hanging around." The psychic said that they did not wish their burial plots to be disturbed and so took to haunting the area.

The men returned again though, and the article reported, "The treasure story was too much of a lure, however, and the men did not stay away long. This time the band of dead men went deeper into their bag of tricks and decided to show the boys just what a ghost really can do."

It began with shouting from every direction, followed by lights in the tree tops. "The woods looked like phosphorus in the moonlight," the paper reported, and added that the "weird lights shone in the ground, in the treasure hole and under the men's feet." When this failed to dispel the men was when the invisible hail came raining down from the sky, pelting the men until they ran away. As they did so, the four farmers swore they heard ghostly laughter. The article concluded that the men had dug only at night so as to keep the treasure a secret, but now that the story was out, they would dig during the daylight hours in front of a crowd to see what happened.

Terror in the Mines!

O.M. Purcell, a former soldier who fought in WWI from Wichita, Kansas, decided to call the bluff of the ghosts and came ready to fight. He offered to do so to the tune of $1,000. His formal offer, published in the papers, claimed he would work the treasure hole for a full 24 hours. If he stayed there and never fled his post, he was to be paid $1,000. If he fled, then he was obligated to pay the farmers $1,000 himself.

World War Veteran Offers Services to Allay Ghosts

The *Okmulgee Daily Times* of February 15, 1927, reported, "Ghosts of Snake Creek Fail to Make Appearance" on their second page as the main headline. Anywhere from 200 to 300 spectators from the region descended on Snake Creek to see the ghosts for themselves, and naturally they had stage fright.

WEEK-END CROWDS VISIT TREASURE SCENE; CRITICAL

World scoffs as Natives Declare Tales Are Really Truth

Despite the ghosts failing to show, Lee stuck to his guns, declaring, "They can laugh all they want to. But there is something down there that mortal man can't produce and I wouldn't go near it again at night for anything you'd offer." In the same article, Lee clarified that a man named Hayworth Baker had told him of the spot before that, and that Baker himself had "been scared away four or five times until he was pretty shaky" himself. "It was new to the rest of us, but his stories had us on the lookout, you can bet," Lee added.

He better described the location of the treasure, stating that it was located at the base of the biggest tree in the area, which contained a hole full of water at its base. It was about 10:30 PM, and as they bailed water out of the hole, they heard what at first sounded like a hoot owl until it became far too loud. "Boys, here she comes," Hayworth said, glad to have a few fellow witnesses with him.

> Pretty soon the long howl was repeated and rose at the volume of an Indian war whoop—a wild, piercing scream. It was rushing up the creek at the speed of an express train. It came up to within a few feet of us and divided into a thousand yells, shouts, cries and everything else. One of them, close at hand, sounded like a woman wailing and weeping. Something kept speeding past our ears going "zzzzt" like a bullet.

The article continued that, "Up on the hilltop a bunch of lights, like balls of fire, began shooting from the ground." The men then ran through the creek while being pelted with the invisible hail. Lee's wife vouched for his story, recounting how fast he had ran home that night and also how terrified he had looked. Lee went on to claim he could produce a list of thirty other witnesses who had been to the spot and experienced supernatural activity.

Despite reports of an earlier unidentified psychic who had ascertained the identity of the six spirits, an area spiritualist with psychic abilities identified as W.W. Purvine investigate the area for a full twelve hours from 11AM to 11PM. In his opinion, there were no spirits there that he could sense.

Purvine pondered that perhaps one of the farmers had psychic abilities themselves which attracted the spirits to the place during the encounter. Purvine investigated the spot with six fellow members of the National Spiritualist Association.

SPIRITUALISTS TO INVESTIGATE GHOST 'SEXTET'

W. W. Purvine Will Head Party to Visit Snake Creek

EXPLAINS SPIRIT MANIFESTATIONS

Earthbound by an Affection for Something Near Spot Reason for Guarding Treasure

The Okmulgee Daily Times (February 12, 1927).

Purvine stated that "Spirits who are earthbound, are held by affection of something that was very dear to them here on earth. If there is gold at the spot described, and it is being guarded, it is probably that the spirits are held there by attraction to it. They will remain there until the spell is broken." This is what Purvine said initially, but upon investigation, it was his opinion that there were no spirits there at all.

9.

THE CREATURES IN
THE COAL MINE

SEPTEMBER / OCTOBER 1903
VAN METER, IOWA

In September and October 1903, about 20 miles west of Des Moines, Iowa, in the tiny town of Van Meter, there arose one of the most bizarre "creature" stories ever. It involved at least two strange, winged beasts, possibly prehistoric or maybe even extraterrestrial, that were seen by multiple witnesses at several different locations. Van Meter had a population of 407 in the 1900 census and seems an unlikely place for a paranormal incident of this magnitude to have occurred. Yet the event is still remembered and celebrated today. The creatures had apparently taken up residence in a coal mine located on the outskirts of town, from where they made occasional forays into the surrounding area, terrifying many of the local townspeople.

Said the October 4 *Des Moines (Iowa) Daily Record*, "The town of Van Meter, containing 1,000 persons is terribly wrought up by what is described as a horrible monster. Every man, woman, and child in the town is in a state of terror, and fully half of them fail to close their eyes in slumber except in broad daylight. Friday night while the population was keeping the dreadful vigil every electric light in the city was kept burning to satisfy the most timid."

The creature, which was later found to actually be two separate beasts, was described in the paper as "half human and half beast, with great bat-like wings. A dazzling light ... came from a blunt, horn-like protuberance in the middle of the animal's forehead, and it gave off a stupefying odor."

Abandoned Mine. Famartin / CC BY-SA (https://creativecommons.org/licenses/by-sa/3.0)

After the creatures were seen at several locations in the vicinity, their lair was finally discovered. Early on the morning of October 3, a worker at a nearby tile and brick factory, J. L. Platt, Jr. heard strange noises coming from the vicinity of the coal mine and went to investigate. The sounds led him to where the bizarre creature was lurking, accompanied by a second, somewhat smaller beast of the same type!

Seeing Platt, the creatures ran away, Platt headed into town to let everyone know that he had discovered the lair of some type of horrible monsters. A posse was assembled, and it headed down to the mine to await the appearance of the creatures.

At around dawn, the armed posse saw the two-winged monstrosities approaching the mine, and the men opened fire. Unfazed by the hail of bullets, the creatures easily moved past the men and made their way deep into the old mine.

But at this point in the story, let us digress to a number of very impressive sightings of the creatures in the days prior to the discovery of their lair. The incidents began when Van Meter resident Uly Griffith, a traveling tool salesman, returned to his hometown after midnight on September 29, 1903. He noticed a strange light atop one of the neighboring buildings, which he thought was suspicious. Walking closer to the light, he saw that it moved with amazing speed to another roof across the street. Puzzled, Griffith then noticed that the mysterious light suddenly disappeared altogether.

The next incident happened about 24 hours later, early in the morning of September 30. The town physician, Dr. A. C. Olcott, was awakened by a bright light shining in his face. Grabbing his gun and going outside to look for the source of the light, Olcott saw what looked like a half-human, half-animal creature with "great bat-like wings" and a "single blunt horn" on the forehead. It became clear to Olcott that the bright light was coming from the creature's horn.

Shocked and frightened, Olcott raised his weapon and fired five times, striking the creature. Shockingly, the bullets had absolutely no effect on the beast. Alcott quickly retreated into his office, locking the door securely behind him.

Sketch of Creature (Courtesy Van Meter Public Library).

The creature made another appearance during the early morning hours of October 1, when it was seen by Clarence "Peter" Dunn, who worked at the town's bank and watched the bank building at night. Armed with a shotgun, he was soon alerted to strange gurgling noises outside the bank. Through the front window, he saw a very bright light shining upon him, which he later saw was coming from the bizarre creature. He fired the shotgun, shattering the glass of the front window, but the creature disappeared. He later discovered large, three-toed tracks where the creature had stepped in the soft earth.

DES MOINES' NEW MONSTER

Citizens Tell Weird Tales of This Modern Terror.

Des Moines, Oct. 12.—According to prominent citizens, two weird-looking, terror-striking monsters are living in an abandoned coal mine on the edge of the town. At night they come out and act as a sort of town curfew bell—every one locks the doors and hides under beds or behind curtains. Residents whose veracity heretofore has been unquestioned tell harrowing stories of experiences with the horrible monsters.

Dr. A. C. Olcott, awakened by a bright light shining through his window, says the terror he saw was half human and half beast, with great bat-like wings. A dazzling light that fairly blinded him came from a blunt, horn-like protuberance in the middle of the animal's forehead, and it gave off a stupefying odor that almost overcame him.

Peter Dunn, cashier of the bank, fired his shotgun at the monster. Next morning imprints of great three-toed feet were discernible in the soft earth. Plaster casts of them were taken.

Dr. O. V. White saw the monster climbing down a telephone pole, using a beak much in the manner of a parrot. As it struck the ground it seemed to travel in leaps, featherless wings to assist. It gave off no light. He fired at it, and he believes he wounded it. The shot was followed by an overpowering odor.

Sidney Gregg, attracted by the shot, saw the monster flying away.

J. L. Pratt, foreman of the brick plant, heard a peculiar sound in an abandoned coal mine. Presently the monster emerged from the shaft, accompanied by a smaller one. A score of shots were fired without effect.

The whole town was aroused and just at dawn the two monsters returned and disappeared down the shaft.

The Courier of Waterloo, Iowa
Oct. 12, 1903, p. 1

Late that same evening, the creature was spotted once more, this time by O. V. White, who kept lodgings in an upstairs room of the hardware and furniture store that he co-owned. Grabbing his gun, White soon located the mysterious creature crouching on the crossbar of a nearby telephone pole. White fired at the creature and thought he had wounded it, but with little apparent effect. Interestingly, White said the creature emitted an odor or vapor that seemed to "stupefy" him. A variation of one of the articles also seemed to imply that White had trouble remembering events, as it said that after being disorientated by the odor that "he remembered no more about it." This statement is open to interpretation, of course, but ufologists could associate this statement with the missing time phenomena common to many abductees.

Train Station in Van Meter, Iowa, in 1907, Olson Photograph Co. (Plattsmouth, Neb.) / CC BY-SA (https://creativecommons.org/licenses/by-sa/4.0)

The sound of White's gunshot brought another local storeowner, Sidney Gregg, out to see what was happening. Gregg saw the creature, which he described as being at least eight feet tall with the light on its horn being as bright as an "electric headlight." Gregg said the creature flapped its wings and leaped like a kangaroo, at one point standing erect on two feet before dropping down to all fours and springing away. It then took flight with its wings, seemingly headed to an abandoned coal mine at the edge of town.

It was in the coal mine that an armed posse headed on October 3 after two of the creatures were spotted going into the mine by J. L. Platt. The posse spotted the two-winged creatures and opened fire, but the monsters ignored the bullets and moved deeper into the mine, where the men were unwilling to follow.

The story ends with the townspeople deciding to barricade the mine entrance and trap the unearthly beasts inside. No further information has been found about what happened in the end, including whether the barricade idea worked. Apparently, the creatures were never seen again.

Some UFO researchers over the years have wondered if the "monsters" may have actually been humanoids wearing some type of special suit, such as an astronaut might wear. Is it possible that the extremely bright light coming from the "horn" in the middle of the forehead was really a head lamp attached seamlessly to some type of helmet?

What if these creatures had landed in a spaceship nearby and were exploring the area wearing specialized gear to help them cope with the unknown atmosphere?

If not extraterrestrials, then perhaps the creatures were some type of pterodactyl-type animal. First, newspaper articles mention a professor (presumably a well-educated man) in town stating that he believed it was an antediluvian monster. That the monsters were spotted in a pair, presumably a mated male and female, or a parent and its offspring, also lends some credence to the story. Usually, monsters were just "one-offs" in old newspaper articles, and it was fairly rare for a pair to be spotted.

The odor emitted from the monster is one of the oddest details of all, and actually matches other pterodactyl sightings reported over the years. What's even stranger is that the odor knocked one man unconscious and he remembered nothing after that point—almost like missing time in a UFO case.

The featherless wings would match those of a pterodactyl, as does the "protuberance" on the head (presumably the beak?). The fact that the creature emitted light actually doesn't knock it out of the realm of credibility—well, credibility as far as remnant pterosaur sightings go. There are more than a few cases where the creatures are seen to be bioluminescent. A college biology professor, Peter Beach, claimed to see bioluminescent pterosaurs over a river in Washington State. A woman on a cruise ship in the Caribbean also saw what she described as glowing pterosaurs flying over the water. Pterosaur researcher Jonathan David Whitcomb has even found reports of glowing ropens, a variety of African pterosaurs, from Umboi Island were the witness described the creature as glowing red, like the embers of a fire.

And, if said creatures had an underground nest then it would make sense that they have bioluminescent abilities. Speaking of pterodactyls living in mines, there are other stories to corroborate this. In 1856 the *Illustrated London News* ran an article about French miners discovering a pterodactyl in an underground tunnel. However, that article is very much believed to be a hoax, which is unfortunately what some people think of the Van Meter Monster.

Today the Van Meter Monster incident is fairly well remembered, and a lengthy book on the incident was written for those who would like to research it more: *The Van Meter Visitor: A True and Mysterious Encounter with the Unknown* by Chad Lewis.

Lewis did serious research on the case, even camping outside the famous mineshaft one night. He also investigated the witnesses and found that they were respectable town citizens. "These weren't town drunks," Lewis told the *Des Moines Register* in 2013. The town librarian, Jolena Walker, also had good things to tell the *Des Moines Register* about the men. "Those guys wouldn't have wanted that publicity," she said. Walker also visited the area of the old mine and got an uneasy feeling. So too did the owner of the land, John Jungman.

Like Roswell, New Mexico's UFO Festival, in 2018 Van Meter held its first-ever Van Meter Visitor Festival. Walker told John LeMay that same year that, "We have many more citizens who embrace the legend. We do have few who don't think it was anything, or that nothing happened, but some of them are from families of original settlers. We get the feeling they try to hide it or not discuss it."

Whether you think of the strange Van Meter Monsters as aliens, dinosaurs, or just plain monsters of the wildest type, the story is undeniably a good one.

Le Vampire, lithographie de R. de Moraine, tirée des *Tribunaux secrets*.

1864 French lithograph showing villagers opening a grave to destroy an alleged vampire.

Simpson Mine in Lafayette, Colorado, 1888-1920.
(Courtesy Boulder Historical Society/Museum of Boulder)

10.
THE MINER WAS A VAMPIRE?

EARLY 1900S
LAFAYETTE, COLORADO

In the late 1800s and early 1900s, the small Colorado town of Lafayette, a mere twelve miles east of Boulder, had become a booming mining town. Men came from miles around, and even from overseas, to try their hand at mining, in hopes of gainful employment. Historical records indicate that among those who came to Lafayette were two miners from Europe, 37-year-old Todor "Theodore" Glava and 23-year-old John Trandafir, who found employment at the Simpson Coal Mine in Lafayette. Coincidentally, both men had ties to the mysterious region of Romania known as Transylvania, where Glava had been born and where Trandafir's home was.

Transylvania, of course, was the setting of Bram Stoker's famous 1897 novel about vampires, *Dracula*, which popularized the notion that supernatural creatures exist which feast on the blood of the living to maintain their "undead" existence. Stoker's novel drew heavily upon the many myths and legends from the Transylvania region, as well as other parts of the world.

Of the two miners from Transylvania, one in particular, Todor Glava, has been suspected by some of having been an actual vampire. Although Glava made his home in Austria, he was a native of Transylvania. Both Glava and Trandafir died in December 1918, victims of the worldwide Spanish flu pandemic of that time. They were both buried in the Lafayette town cemetery, where they remain interred to this very day.

Theodore Glava

Theodore Glava, an Austrian miner employed at the Simpson mine, died suddenly Wednesday morning, following an attack of influenza. He had so far recovered as to be up town the evening before his death, but suffered a relapse. He was aged 43 years and is survived by his wife, who is in Austria. Burial will be made this afternoon.

JOHN TRANDAFIR

John Trandafir, a native of Rumania, and an employee of the Simpson mine. died of pneumonia at St. Joseph's hospital Wednesday afternoon of last week, Death came a few hours after he reached the hospital. The deceased was aged 27 years and had been in Lafayette about four years. He is survived by his parents and three brothers, who reside in Rumania. He was a member of the Orthodox Greek church, a n d funeral services were held under the auspices of that church last Sunday afternoon, conducted by the Rev. A. Kaimakan, of Denver. Music was furnished by the City Park band of Denver, and the remains were laid to rest in the Lafayette cemetery.

From the Lafayette Leader newspaper. Glava's death notice appeared on 12-6-1918, and Trandafir's obituary appeared on 12-13-1918.

The local newspaper, the *Lafayette Leader*, in reporting on their deaths, stated that the men had arrived from Europe four years earlier. They had both worked at the Simpson Mine in Lafayette, which opened in 1888 and is believed to have been the first of many mines in the Lafayette area. Glava's first name of Todor is given in its Americanized equivalent of Theodore. Glava is said to have been a victim of the Spanish flu, and Trandafir, whose cause of death was pneumonia, was almost certainly also a victim of the pandemic that was raging throughout the country and the world.

The belief that Glava was a vampire began after his death, when the man's grave supposedly became a hotbed of paranormal activity. The locals noticed that a tree grew right out of the part of Glava's burial site that would correspond to the man's upper torso. As the tree grew larger and larger, so did the legend that at some point, a wooden stake had been driven through Glava's heart and from that piece of wood had grown the tree.

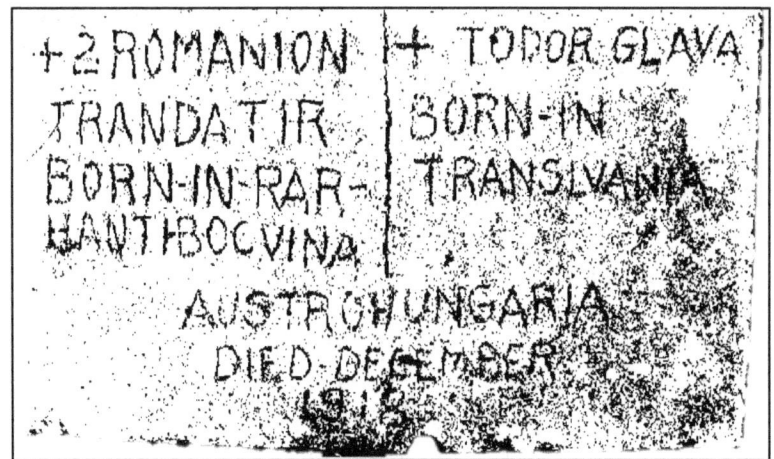

Headstone marking the graves of Both Glava and Trandafir.

The theory was that if someone found it necessary to pound a stake into Glava's heart, this surely indicated that he had been a vampire – in the traditional, mythic sense of the word. According to the legend, possibly due to stories of strange happenings near the man's grave, the townsfolk dug up Glava's remains shortly after his burial. They discovered that his corpse had fresh blood on its mouth, and that its teeth seemed much longer than a normal person's. Also, despite being dead, Glava's hands had grown extremely long fingernails! Undoubtedly, these strange circumstances led the townsfolk to hammer the aforementioned stake through Glava's heart, ensuring that the vampire would cease to exist and eventually causing a tree to rise out of the grave and stand sentinel over the dead vampire's remains. Or at least, this is one theory. Another is that the vampire legend began after the local people noticed the tree growing from the grave.

Asked for her opinion about the legend, Claudia Lund, curator of the Lafayette Miner's Museum, said that the story no doubt resulted from the association of Transylvania with Count Dracula and vampirism. According to Lund, because Glava was apparently from Transylvania, there was some uneasiness about his status after death.

To this day, people claim that the grave is the site of much paranormal activity with spectral figures seen hanging about along with disembodied voices and strange lights. Other accounts claim that all battery-operated devices are quickly drained of electricity when brough into close proximity with Glava's grave. People have also claimed to have been attacked by a mysterious figure, and the only clues are a set of footprints leading back to the grave.

As it stands, there never were any stories of either Trandafir or Glava attacking the populace of Lafayette, either before or after their deaths. It's possible that the vampire legend began after townspeople noticed a tree growing from the spot where Glava's torso would have been. "A tree, supposedly, mysteriously grew up from the grave where his heart would have been, and people wondered if there wasn't a stake driven through his heart because they thought he was a vampire," Claudia Lund explained. When blood-red roses also grew near the grave (said to be the vampire's fingernails), that further amplified the legend.

After the myth was established, local children would dare each other to stand upon the alleged vampire's grave. Then, stories started to emerge about ghosts being sighted near the grave, along with strange voices being heard. A "ghost hunter," identified as Anam Paranormal, took an EVP (electronic voice phenomena) device to record the noises near the grave and also an EMF device, which detects electromagnetic fields. The results were surprising.

Anam Paranormal said, "EMF ranged from zeros to maxes and never really stayed at one reading or another the whole time, so good luck getting a decent base reading for comparison. Even in the daytime, EMF went too nutty." Spookier than that, the EVP recorded a voice stating, "You want my stake?"

However, Anam Paranormal doesn't believe that the spirit residing near the grave is an actual vampire, but rather a disembodied spirit "playing" off the grave's reputation - an interesting theory for certain.

11.

NAVAJO GHOST HAUNTS MINE

1880s
NEAR LAS CRUCES, NEW MEXICO

The *Lincoln Daily Nebraska State Journal*, on October 10, 1892, ran a fascinating article about a New Mexico silver mine that was supposedly haunted by the ghost of a dead Native American of the Navajo nation. The article, which was subsequently reprinted in many other papers, quotes several witnesses who claimed they saw the spectral image of the departed Navajo standing about near the entrance to the mine, apparently determined not to allow anyone to plunder its treasure.

The silver mine in question is located in the "Santiago Mountains" of New Mexico, near the city of Las Cruces. This mine reportedly had a legend associated with it that was well known to the surrounding residents. According to the legend, about 50 years prior, a Navajo chief named Jacopo had become aware of the location of the silver mine, but he would not disclose the information to a group of white prospectors that demanded to know where the mine was. When their efforts to get Jacopo to speak were unsuccessful, the white men killed the Navajo chief.

For this reason, the story goes, Jacopo's spirit appears near the site of his death and functions as a guard to keep trespassers away from the silver ore inside the mine. Residents who live and work nearby are very careful to avoid the area, for fear of running into the phantasm.

SEEING THE SPECTER.

Newspaper Illustration of the Ghostly Encounter.

About fifty years after Jacopo's demise, his ghostly form was first spotted by two young Americans from Milwaukee, Henry Williams and George Goggans, engaged in hunting and fishing along the Madre de Dios River, which flows through the Santiago Valley. One night while camping near the river, Goggans awoke from a frightening dream and told Williams, "I dreamed repeatedly last night of an old Indian who, in full war dress, walked all around the camp."

Amazed, Williams told him that he also had dreamed the very same thing!

A couple of nights later, Williams saw the figure of the Navajo chief standing right behind Goggans and warned him. The moment Goggans turned to see the figure, it vanished from sight.

The following morning, the two men were having breakfast when they noticed the same eerie figure watching them from a nearby clump of timber. They both broke into a run toward the figure, but it "disappeared like a puff of smoke."

That night, they both dreamed that the Navajo had appeared and warned them to confine themselves just to their hunting and to stay away from "the silver mine," which they had not previously heard anything about. Upon waking, the men resolved to search the area in search of the mine.

However, while scaling the mountainside, Goggans slipped on a rock and hurt his foot. As he lay groaning from the pain, the phantasm appeared again, pointed to Goggans' foot and gave a series of grunts, as if to express derision.

Convinced that the injury had happened because of the ghost's displeasure, the two men ended their trip and went back into town.

Among a number of other witnesses of the apparition mentioned in the newspaper article are a "Colonel Jenkins" and Fred Lathrop, "both prominent citizens of this place, and both of whom have seen the phantom several times."

As far as Jenkins' encounter, he reported that he had obtained a hand-drawn, partial map from a Native American of the Zuni tribe who passed it to him as the Zuni lay dying. According to Jenkins, the native had "lost his life in looking for the place."

With the map in hand, Jenkins engaged the services of a gentleman from Tennessee named Wat Houston, whom Jenkins chose partly because "he had no fear of ghosts." Jenkins had heard the legend about the haunted mine and did not want his companion to run off in fear. In 1873, the two men headed out in search of the mine.

Shortly after arriving in the general area, the men were riding down a gorge when they suddenly saw the figure of the Navajo chief standing about twenty feet ahead of them. "He was drawn up to his full height and was in full wardress, Eagle feathers, war paint, etcetera, and was eyeing us from under his bent brows, as if trying to make us out. My horse caught sight of him at about the same time and began to rear and plunge in such a manner as to preclude all progress on my part, but Houston rode on...."

(Library Of Congress).

Jenkins watched his companion approach the spectral figure, when he saw Houston suddenly "shoot out of the saddle over the horse's head and fall in the road ... perfectly motionless."

As Jenkins moved toward the prostrate form of his companion, the Navajo ghost vanished before his eyes. Dismounting, Jenkins checked the limp form on the ground before him and discovered that Houston was dead from a broken neck.

"I put the body on his horse and led him back to town. So ended for the time my attempt to find the mine, though I have made many since then. Each time, however, I see that ghostly

Indian, and the expedition invariably has cost me dearly in some way. Once, I attempted to ride the figure down, and it was like trying to ride through an iceberg. And I can tell you, I would not repeat the experiment for all the silver in the lost mine he guards."

For his part, Fred Lathrop told about seeing the ghost while traveling across the Santiago Mountains to visit a friend's ranch. "I was jogging along leisurely, when raising my head, I saw an Indian in the road directly in front of me. He was apparently of great age and was resplendent in war paint and feathers. I was struck by the oddity of the figure. I hailed him. He waved me back, exclaiming something in his native tongue or Spanish, but what I do not know. I continued to advance, however, and was almost on him, when my horse refused to go another step."

The Santiago Mountains in Texas
Rebelcry, CC BY-SA 3.0 <https://creativecommons.org/licenses/by-sa/3.0>

Lathrop then noticed something very odd. "I then perceived to my astonishment that I was able to see objects in the road beyond him, just as if he had not been there, though the figure, when looked at closely, seemed as palpable as any other thing about me. I paused, scared nearly to death, as I do not mind confessing to you, and too much bewildered to even turn my horse around."

85

The ghostly figure came forward and took the horse's bridle in his hand, leading it all around the clearing. "While it was being held by the ghost, the poor horse kept startling and was bathed in perspiration."

The phantom continued to lead the horse "for a space of 100 yards," while Lathrop sat frozen with terror, unable to resist or protest. "When we had reached a road that circled the mountains without passing over them, the ghost gave me a farewell wave of the hand, and was gone as an image from a mirror."

Lathrop concluded his narrative by saying, "Yes, I was a skeptic on the subject of spirits returning to the scenes of their mortal lives, but since that experience, I have never doubted that there is at least one ghost."

This story, as intriguing as it sounds, has some question marks associated with it. In modern times, there are no "Santiago Mountains" located near Las Cruces, New Mexico. It is uncertain if any of the mountains around Las Cruces may have, at some point in past history, been known as the Santiagos. About 300 miles southeast of Las Cruces are the Santiago Mountains of West Texas, although it would be a stretch to say that these are "near Las Cruces."

Similarly, there is no "Madre de Dios" river near Las Cruces, either. The only river by that name is in South America, running between Boliva and Peru. It is uncertain if any U.S. body of water has ever been known as "Madre de Dios." So geographically speaking, this description of the site where all this happened seems problematic.

As for the individuals named in the story, we did find genealogical data for a number of persons with the names of George Goggans, which is somewhat of an unusual name, and also Fred Lathrop. However, since few other details about them were given, it is difficult to pinpoint their identities with certainty.

The authors must conclude that there is insufficient historical evidence to verify this story as true, but nonetheless it is quite fascinating.

12.

WRAITH IN A TOMBSTONE MINE

1890s
TOMBSTONE, ARIZONA

The city of Tombstone, Arizona, is well known in history for its legacy of gunslingers, outlaws, lawmen, lynching, brothels, saloons, and gunfights. But in 1897, a strange story arose about a "haunted" mine just three miles outside town, known as the Bronco Mine. "According to the stories told by eyewitnesses, two hours after sunset with clock-like regularity, a tall, white or luminous, wraith-like form stalks about among the diggings and passes through the old adobe shanty near the mouth of the mine shaft," said the *San Francisco Call* on May 27, 1897.

"The rugged mountaineers and plainsmen of the area, who are not a superstitious lot, reportedly have, on numerous occasions, fired their weapons at the specter, sometimes at very close range. They have also chased the entity, trying to corner him, but upon approaching the creature, it vanished and then reappeared a short distance away from them," it continued.

Oddly, the apparition apparently had a strong work ethic and was observed carrying out mining "work" deep in the abandoned mine, despite the fact that it had long been depleted of its silver deposits. "For hours he has been heard at work in the deserted drifts, now pounding drills, now sawing timbers, now blasting. He works along as industriously as though silver had never depreciated, and his labors are so unceasing that half the population have heard them and really believe that the weird sounds and sights are genuine."

Entrance to Tombstone. (NYC Public Library)

Entrance to early silver mine in Tombstone area (Library of Congress)

Terror in the Mines!

To fully understand this ghostly apparition, one must consider the Bronco mine's history of violence and murder. In its early days, the Bronco was one of the richest mines of the Tombstone district. Over the years, as massive profits were reeled in, arguments and fights arose among the mine's owners. Occasionally, an owner or employee would suddenly "disappear," never to be seen again – presumably a victim of the violence and greed engendered by the mine's riches.

There were reports of many shootings at the mine and surrounding area. At one point in the early 1880s, five men were found murdered there. Later, the area of the Bronco mine became a base of operations for gangs of thieves and stage robbers. In one case, a gang who robbed a Wells-Fargo bullion wagon brought their booty back to the spot, after which a violent argument ensued over the division of the spoils, resulting in the deaths of the entire gang – to the last man.

Tombstone Mining Operation, Circa 1880 (Lib. of Congress)

Many killings occurred at the infamous Brunckow Cabin, located near the entrance to the mine. According to *Geocaching.com*, "In early days of the Tombstone claim, the Brunckow Cabin was the scene mayhem and carnage." It explained that numerous shootings occurred there and that men would come up missing. For instance, one man was supposed to have been shot and thrown down a well. However, killings and disappearances were so common, it wasn't uncommon for them to go uninvestigated. Charleston Road in particular served as a rendezvous point for assorted bandits. It was there that five corpses were found that had recently robbed a Wells Fargo bullion wagon. The men had apparently fought and killed one another while arguing over who got what. Eventually, a US Marshall moved into Brunckow Cabin and was murdered only a few weeks later. According to historians, about 21 men died either in or near the cabin. "Rumor has at least three graves in the area, to include the final resting place of Fredrick Brunckow himself," according to *Geocaching.com*.

HAUNTED ARIZONA MINE.

A weird ghost story comes from the old Bronco mine, three miles southeast of Tombstone, Arizona. Many citizens, otherwise entirely truthful and reliable, relate wonderful tales of the strange sights and sounds that occur nightly in the old mine, and these have been repeated with such insistence and with such circumspection of detail that at last people have ceased to scoff and sneer, and an investigating party is being made up of volunteers to go down into the mine and pass a night at the bottom of the shaft.

According to the stories told by eyewitnesses, every night a tall, white or luminous form stalks through the old adobe shanty near the mouth of the main shaft. At midnight he ceases his meanderings on the upper levels and goes below to work, now pounding drills, now sawing timbers, now blasting. The Broncho mine has a bloody history that well entitles it to its ghostly tenant. Shooting affairs were numerous at the mine over the division of profits, and at one time in the 80's five men were found dead there.

Sierra County Advocate
(Hillsboro, NM), 6-4-1897, page 3.

The *Tombstone Epitaph* printed a ghost story featuring Brunckow Cabin as early as May of 1897, which is what the earlier *San Francisco Call* article was based upon, and mentioned the haunted mine along with it. According to the article, every night "a menacing ghost was seen stalking around and through the dilapidated adobe shanty." The curious and the brave reported that when they got near enough to speak t the apparition it always vanished only to reappear instantly at another point. "Even today, ghost hunters and paranormal investigators flock to the Tombstone area to have a chance of investigating at the legendary haunted Brunckow Cabin," concluded *Geocaching.com*.

Terror in the Mines!

The *San Francisco Call* article was based on a shorter piece that ran four days earlier in the *Tombstone Epitaph*, which read: "The halcyon days of Tombstone are often brought to memory, and if even at the expense of some unfortunate, it is a pleasure to allow the mind to revert back to the 80s when this was surely the greatest mining camp on earth; when the shrill whistles of the numerous mines were deafening to the ear; when the bad man prevailed and the music of his guns lulled many to sleep; when bold highwaymen plied their vocation almost unmolested; when 'life' began with the fading away of each day and there was one continual round of pleasure. Crime then was regarded as a matter of course, criminals held full sway and it was a case of survival of the fittest. These reflections are brought to mind through the report of a spook which has taken possession of a long-since abandoned mine, the history of which would at once establish it as the appropriate habitation of ghosts. In early days the Bronco mine, three miles below Tombstone, was the scene of much excitement; dissension arose among the owners and shooting affairs became numerous, occasionally a man was missing and that ended it; one man was supposed to have been shot and thrown into a well, but as there was an abundant man in those days an investigation was deemed unnecessary. Five men were found at the Bronco with their toes pointing heavenward at one time; it was a local rendezvous for the knight of the road, and the five dead men found there were a part of freebooters who had raided a Wells-Fargo bullion wagon and fought over a division of the spoils.

"If such a thing be possible, then it is no wonder that the spirit of the departed should linger around the scene of pillage and carnage, and reputable men of Tombstone will vouch for the truthfulness of the statement that the mine is haunted. The story goes that every night can be seen a menacing ghost stalking around and through the dilapidated adobe shanty; people have attempted to investigate, but upon approaching apparently near enough to speak, the spook suddenly vanishes, only to appear as quickly at some other point, leading its would-be quarries on a lively and elusive chase. There is apparently but one, and when not to be seen mining operations can be heard in the old shaft, pounding on drills, sawing lumber, and working along ever and anon just as though silver had never depreciated. That there is some mysterious movements around the Bronco is honestly believed by many here, and mountaineers will visit this deserted mine and attempt an investigation."

SAW A GHOST AT WORK IN A MINE

Strange Stories Told by Reputable Men of Tombstone.

White-Robed Figure Which Has Proved Impervious to Bullets.

May Be a Device to Scare the Inquisitive Away From a Rich Find.

TOMBSTONE, Ariz., May 26.—A weird ghost story comes from the old Bronco mine, three miles southeast of here. Many citizens, supposedly truthful and reliable, relate wonderful tales of the strange sights and sounds nightly in the old mine, and these have been repeated with such insistence and with such circumspection of detail that at last people have ceased to scoff and sneer, and an investigating party is being made up of volunteers to go down into the mine and pass a night at the bottom of the shaft.

According to the stories told by eyewitnesses, two hours after sunset, with clock-like regularity, a tall, white or luminous, wraith-like form stalks about among the diggings and passes through the old adobe shanty near the mouth of the main shaft. Many of the mountaineers and plainsmen of the neighborhood are without superstition and these have tried to catch the specter. They have shot at him time and again, and often from very close range. They have tried to corner him and sense his substance by material touch, but always he has vanished at the critical moment, only to reappear at a little distance.

At midnight he ceases his wanderings on the upper levels and goes below and to work. For hours he has been heard at work in the deserted drifts, now pounding drills, now sawing timbers, now blasting. He works along as industriously as though silver had never depreciated, and his labors are so unceasing that half the population have heard them and really believe that the weird sounds and sights are genuine.

The Bronco mine has a bloody history, that well entitles it to its ghostly tenant. In the early days it was one of the richest mines of the Tombstone district. Dissensions over the division of the profits arose among its owners, and occasionally, as these quarrels ripened, a man or two disappeared. Shooting affairs were numerous at the mine. At one time in the early '80's five men were found at the Bronco with their toes pointing skyward. Later on the mine became an ideal rendezvous for the freebooters and stage-robbers, and once a gang of these who had robbed a Wells-Fargo bullion wagon fought over a division of the spoils and killed each other—to the last man.

One reasonable explanation is offered to account for the appearance of the supposed wraith in the mine. It is that some one who is acquainted with the Bronco has found a pocket of rich ore, which he is working at night, while, in order that he may remain unmolested, he dons ghostly raiment, calculated to keep the inquisitive at their distance.

*San Francisco Call,
May 27, 1897, Page 3.*

As stated before, many people believed that the area was haunted by the spirit of one of the many men who had died there. But among the less superstitious, a different theory arose, suggesting that "someone who is acquainted with the Bronco has found a pocket of rich ore, which he is working at night, while, in order that he may remain unmolested, he dons ghostly raiment, calculated to keep the inquisitive at their distance."

Terror in the Mines!

Nonetheless, there apparently were attempts to get to the bottom of the mining mystery, as reported in the June 4, 1897 edition of the New Mexico newspaper *Sierra County Advocate*, which said, "Many citizens, otherwise entirely truthful and reliable, relate wonderful tales of the strange sights and sounds that occur nightly in the old mine, and these have been repeated with such insistence and with such circumspection of detail that at last people have ceased to scoff and sneer, and an investigating party is being made up of volunteers to go down into the mine and pass a night at the bottom of the shaft."

Unfortunately, no historical record has been found of whether this investigation was ever conducted or what its results may have been.

Ghosts Give Miners Warning.

FORT DODGE, Dec. 11.—Strange stories of spooky sights and sounds come from the Collins Bros.' coal mine, at Coalville, in the southern part of this county. The miners say that ghostly rappings, whistlings and the sound of unseen picks are heard by the men on the night shift. James Grant, one of the night miners, solemnly states that he saw with his own eyes the disembodied spirit of another miner warning him not to stay in that part of the mine. He at once threw down his pick and could not be induced to enter the haunted room again. The miners think that some great danger threatens them and go to work each day with fear and trembling.

Winona (Minnesota) Daily Republican (December 11, 1890, page 4).

13.
SPECTER IN A
COAL MINE

1890
COALVILLE, IOWA

In December 1890, a strange episode occurred at a coal mine in the Coalville – Kalo area, Iowa, located six miles south of Fort Dodge and about 90 miles north of Des Moines. Coal miners working the night shift for the Collins Brothers company began complaining about ghostly rapping, whistling, and the sound of unseen picks heard in the mine during the overnight hours.

One of the night shift miners, James Grant, told newspaper reporters that he saw with his own eyes the disembodied spirit of an unknown miner apparently trying to get his attention. Grant said that he felt the strange apparition was trying to warn him not to stay in that specific part of the mine. Grant became fearful that it was a premonition of a mining disaster that would soon occur.

Throwing down his mining pick, Grant fled the area and refused to enter the "haunted room" again. Subsequent newspaper reports about the incidents said, "The miners think that some great danger threatens them and go to work each day with fear and trembling."

A few months later, in April 1891, a mysterious, catastrophic fire struck at the Collins mine. Newspaper accounts said that all of the workmen in the mine were driven out by the gas and smoke. All attempts to get the fire under control were to no avail. "The origin of the fire is a mystery," said reports.

A Burning Coal Mine.

FORT DODGE, April 27 —A disastrous fire is in progress in the Collins coal mine at Coatville, and the workmen have been driven from the mine by the gas and smoke, and all attempts to get at the flames have proven fruitless. The origin of the fire is a mystery.

Dubuque (Iowa) Daily Herald (April 28, 1891, page 1).

A report filed on April 29 said, "The fire broke out in the mine several days ago, but was thought to have been subdued until yesterday when the flames broke out again worse than ever. All mining operations have been suspended and efforts are being made to check the progress of the fire."

Coal Miner, Circa 1923. (Library of Congress)

Terror in the Mines!

Is it possible that the appearance of the "ghost" in December 1890 was a premonition of impending disaster? Following the inferno that roared through the mine in April 1891, a number of other disasters struck the Collins mine in the ensuing years, including a massive cave-in that killed one miner and injured others. Newspaper reports from January 12, 1893 said, "John Means, a young man employed in the Collins coal mines near here [Fort Dodge], was killed by a cave-in yesterday. He had been at work in the mines but a short time." Genealogical records indicate that the victim's full name was John Thomas Means, born in 1856 and residing in Coalville. He was 37 at the time of his death.

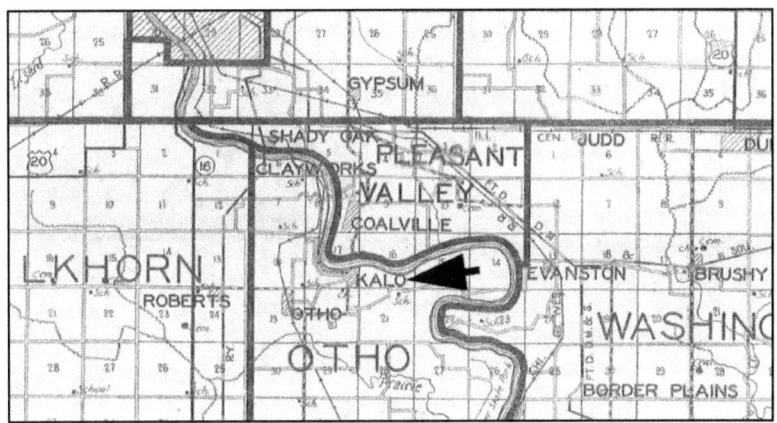

Map of Webster County, Iowa, showing the location of Kalo.

The story that a "ghost" seemed to be warning of disaster is unique among "haunted mine" tales. Many stories of this type involve strange noises, wind, etc. but not ghostly-looking humans. In this case, the witness said he definitely saw a humanoid form, that it appeared to be a miner, and that it seemed to be trying to warn him.

The mine, operated by the Collins Brother company, was known as the Collins mine. It operated for ten years, according to *History of Coal Mining in Iowa* by James H. Lees. "About the same time that the Fort Dodge Coal Co. began operations, the Craig Coal Co. opened a number of mines at Kalo, opposite the river from Coalville. All of these early mines were drifts and relied on natural ventilation. In 1880 the Minneapolis and Saint Louis railroad was built down the river and gave the mines an outlet to the north. By 1883 these two companies were operating six mines and were putting out 600 tons daily. The Fort Dodge Coal Co. was employing 350 miners who dug thirty cars of coal each day. Other early producers were Collins Brothers, who operated a large shipping mine for ten years, and the Standard Coal Co., whose mines were for a time the largest producers at Kalo."

Regarding the witness that originally saw the spectral miner back in 1890, a search of historical records found many persons named James Grant in Iowa during the time frame of this story. No further information could be found on what happened to him after this initial incident.

14.

A SWINDLER, A GHOST, AND A GOLD MINE

1890S

CRIPPLE CREEK, COLORADO

I In 1896, a story was told by old-time miners in Cripple Creek, Colorado, of a particular mine that was rich in gold ore but lay abandoned with nobody willing to exploit its riches due to reports that it was haunted by the ghost of the murdered mine owner. Located at the base of one of the nearby mountains, which was known at that time as "Ball Mountain," the gold mine was known for eerie manifestations seen late at night by people passing by its entrance. Witnesses reported seeing a "white cloud" coming out of the mouth of the shaft, followed by the appearance of the spectral figure of a man, whose head was disfigured by a caved in skull and who was bleeding profusely. If the onlookers tried to move closer to the figure, it slowly sank back down into the mine.

One veteran miner named 'Cheyenne Bill' said that he'd be willing to take over the mine except for the frequent appearances of the ghost of the murdered man. Reiterating that he was not afraid of "anything that runs or walks," Cheyenne said, "If I was sure that the dead man was not in the mine, I'd go there and jump the claim. Oh no, I'm not afraid of ghosts, but I don't want to take advantage of a dead man – that's all."

So, what might have caused the dead, bleeding man with the caved in skull to start appearing at this Cripple Creek gold mine? According to the local miners, the story began when a swindler calling himself "Captain Jack" arrived at Cripple

Creek, finding himself without much money and in great need of lots. Being in the midst of mining country, Captain Jack decided to run one of the oldest mining frauds in the business – bury a few gold nuggets in the ground, persuade someone of great riches, sell them the claim to the mine, and then disappear before the mark realizes the so-called gold mine is worthless.

Cripple Creek, Colorado, Circa 1900. (Library of Congress)

Thus, Captain Jack found an unclaimed spot of land, dug to a depth of twenty or thirty feet, and buried some gold ore several feet down. He then ran his scam on a would-be prospector, visiting Cripple Creek from somewhere in the Eastern United States. The unsuspecting "tenderfoot," whom we will call "Smith," went out to the spot where Jack had buried the ore and started digging where the swindler suggested. After digging for about an hour, he dug up the ore, took it to an assayer, and was told that it was genuine gold ore, running about $100 to the ton.

Not being very knowledgeable about mining and eager to begin exploiting the claim, Smith bought into the scam and paid $20,000 to Captain Jack, who took the money and swiftly left town for Denver. Smith hired some local men to excavate, and before they had gotten down fifty feet, they struck good paying ore. A few feet beyond that, they struck a vein of $500 a ton ore.

Word soon spread of Smith's good fortune, and the news reached Captain Jack in Denver, who by then had gambled away most of the $20,000 he had gotten from Smith. Returning to Cripple Creek, Jack asked Smith for a job at the mine, and, because he greatly appreciated Jack having sold him the claim, Smith made him the superintendent of the mine.

Sometime after Jack started as superintendent, he found himself alone with Smith in the mine after all the other miners had left, except for a few men, some distance away, in the transport "cage" waiting to be hoisted up to the surface. They could not see Smith and Jack from their vantage point. As their cage began to ascend to the surface, the men thought they heard "groans" from where Smith and Jack had been working.

Cripple Creek Mining Operation, Circa 1800s. (Library of Congress)

A while later, the cage went back down and brought up only one person – Captain Jack, who told the others that Smith would be remaining down in the mine for quite a while longer. The men thought this sounded strange, but they said nothing. Jack instructed the miners to go ahead on to supper.

The miners never saw the mine owner, Smith, again. Some days later, Jack told the miners that Smith had decided to sell the mine, and he brought a group of men to look it over. While the prospective buyers were examining the mine and doing some exploratory digging, they dug up Smith's body with its skull caved in.

Remembering the groans they had heard shortly after they saw Smith for the last time, alone with Jack, they determined to try to bring Jack to justice. However, Jack had surmised that something was in the wind, suspecting that the men knew something about his crime, and he left town quickly. It was reported that Smith had gone to live with a tribe of Native Americans, and he was never heard from again.

Terror in the Mines!

For those interested, a version of the story, told in the words of Cheyenne Bill, was printed in the February 24, 1896 edition of the *New Haven (Connecticut) Daily Morning Journal and Courier.*

The hotel accommodations on Ball Mountain being very meager, a great man of the miners, after their day's work is done, walk over to "Cripple" (as they call it for short), where, even if the hotels are full, a chair can be had at fifty cents per night. Here they can sit, warmed by a huge fire, snore, and talk all night – telling of their good prospects and how near they came to turning up a bonanza. The distance being only two miles, they naturally prefer to come over to the "city," where they can take in the dance houses and try their luck in some of the numerous gaming houses. This alleged haunted mine is at the base of Ball Mountain, and, though these miners are not 'afraid of anything that runs or walks,' some of them are superstitious and make a wild detour, as a small boy does in passing a graveyard.

"Cheyenne Bill" squirted a three-foot stream of tobacco juice on the red-hot stove, and by unanimous request proceeded to tell the story of the haunted 'prospect.'

"Wall, boys, they say it's haunted, but I've never seen any spirits 'round, and I would not believe it until I see 'em. No, I ain't looked for 'em. The prospect belonged to a tenderfoot from the cast – I forgot his name. A fellow named Captain Jack – I forget which one, there are so many Captain Jacks – sold this claim to the tenderfoot. He worked the old game on the tenderfoot. After he had sunk a shaft some twenty-five or thirty feet, one night he brought over some $100 to the ton ore and planted it a foot or two. Next day he took the tenderfoot to the mine, and after digging about an hour, he dug up the salted stuff. They went to an assayer, and of course he showed that it was genuine gold ore – running about $100 to the ton. Of course, you all know that a streak of ore may be thin or wide, deep or only surface thin, and the tenderfoot was so eager to buy that he did not care to prospect deeper than a couple of feet. Perhaps Captain Jack would not have allowed it any way. You all, maybe, know how it is when a fellow that's hard up wants to sell a mine." And Bill gave a knowing look as he paused to irrigate the stove with another rivulet of tobacco juice.

"Wall," continued Cheyenne Bill, "the tenderfoot gave $20,000 for the prospect."

There was a whistle and exclamations of approbation from several of the half sleeping listeners, and the stove was assaulted from all directions.

"The tenderfoot," continued Cheyenne William, "put some men to work, and before he had got down fifty feet struck good paying ore, and a few feet further he struck a vein of $500 a ton ore.

This last sentence awoke the old miners who have an ear for little else than stories of "pay ore."

"Captain Jack had gone to Denver, and blown in his money in the gambling houses and divers, and was again back on the range prospecting when he heard of this rich strike in the prospect which he had sold to the tenderfoot as a worthless claim. He went to the mine and asked the tenderfoot for a job, and was made superintendent for the good fortune he had given the tenderfoot in selling him the mine. Wall, boys, you know how it sometimes happens. The tenderfoot owner and his superintendent was down in the mine alone sinking a drill to see how far the vein extended. All the men had come up and gone go supper except the two men at the windlass. These men said that just before the signal to haul up was given they hear something like groans. They hauled up the cage, and only Captain Jack was on it. He told the men to go to supper and that he would stay and haul up the boss, who would be in there some time. The men thought it was strange, but said nothing. Sometimes, you know, when a drill is run and good ore is found the secret is kept from the miners, who 'leak' in their talk. But there were no shares for sale in this mine, and it was not necessary to keep a strike a secret, as they do when they want to buy some stock at their own figures.

"Wall, next day the men were put to work sinking a shaft in another direction, and the superintendent gave it out that the boss had gone to Denver to sell his mine to a lot of eastern fellers. This looked likely, and nothing was said. Soon afterward Captain Jack went to Colorado Springs and came back with some men, who, he said, the boss had sent down to look at the prospect. The boss was not along, and they thought this was strange, too.

"They went down in the mine, saw the ore, and took some to the Springs for assay. When the chief man and Captain Jack were at the Springs some of the party came back and went down in the mine to dig a few feet. They had heard of salted mines, and they were a little afraid that a trick was being played on 'em. They dug around, uncovered some loose rock in a cross shaft that had been just started. There they found the body of a man with his skull broke in. The murderer had drove the pick through the back of his head. The men come out and told what they had seen, and then the boys who were at the windlass recalled the groans they thought they heard, and knew why the boss did not come up when the superintended did. The boys were waiting for the superintendent, and would have had some fun hanging him, but the fellers declared the deal off, and

Terror in the Mines!

by the way they acted the superintendent saw they had found out something, and left the diggings and went to live among the Indians. That was before Cripple was on her feet as a big camp, and nobody has jumped the claim yet. It belongs to the dead man, but everybody knows that in passing at night they see a kind of a white cloud at the mouth of the shaft, and a man comes out with his skull broken in and bleeding, and when they go nearer to see what it is, it sinks back into the shaft. No, the body was never taken out. Maybe nobody told the officers, that I know of, but if I was sure that the dead an was not in the mine, I'd go there and jump the claim. Oh, no! I'm not afraid of ghosts, but I don't want to take advantage of a dead man – that's all," concluded Cheyenne Bill, in answer to a number of remarks reflecting upon his ability to face the ghost of Cripple Creek."

View of explosion tunnel at Dawson, NM, in 1923.

15.
TOMMYKNOCKERS OF DAWSON, NEW MEXICO
1903-1923
DAWSON, NM

L ike many other folkloric creatures, the Tommyknockers originally came to the U.S. from Europe. The gnome-like beings inhabited mines and were called Knockers in Cornish and Devon folklore. They were short, only about two feet tall, with slightly oversized heads. Otherwise, they were humanoid and more or less resembled the titular supporting characters of *Snow White and the Seven Dwarves*. Or, that is to say, they were basically mini-miners. For the most part, they were benevolent and warned of oncoming disasters in the mines, hence the knocking that warned underground workers to get back to the surface before a collapse, flood, or other disaster. However, the Knockers also took to mischief by way of stealing or misplacing tools.

It was during the 1820s that Cornish miners began to spread tales of Tommyknockers in the United States in Pennsylvania. Then, during the California Gold Rush of the 1840s, stories of the mine goblins spread to the western territories. One such Western town to have a more distinctive version of the Tommyknockers was Dawson, New Mexico. The burgeoning mining town began via its first settler, John Barkley Dawson, who acquired the land via the Maxwell Land Grant in 1869. Dawson soon discovered the land was rich in coal, and by the onset of the 20th century the town of Dawson was to be a major mining center. It quickly grew to become the largest coal mine in New Mexico at the time. And, with a heavy presence of miners came Dawson's own form of the Tommyknockers.

Depiction of European dwarves.

In *Hispanic Legends from New Mexico*, an old-timer from Dawson gave an account in 1952 of the mine's old days. Specifically, he spoke of the strange, benevolent creatures to inhabit the mines on page 423:

> The smike is a legendary animal living in the mines of Dawson, New Mexico. If an accident was about to happen, the smikes would all run to the back and farthest parts of the mine. They were a danger signal for the men. Smikes would eat only human excrement. They inhaled the poison gases of the mine and exhaled the oxygen. It is said they have guided more than one lost miner out to freedom.

The smikes may or may not have been on duty on September 14, 1903, when a fire broke out in the No. 1 Mine. In spite of several explosions that also occurred, five hundreds miners evacuated the mines to safety and only three perished. Having survived one disaster, Dawson was purchased by the Phelps Dodge Corporation in 1906. The company worked to expand the town into a model mining center and it soon became renown across New Mexico and the West. The town even had its own movie theater eventually, plus other amenities like swimming pools and bowling alleys.

Dawson, NM.

Dawson also boasted a rather large cemetery. Although the Smikes may or may not have guided some miners to safety, not all were so lucky. Many succumbed to the Black Lung, while others fell into pits or were crushed in cave-ins. But still nothing of the magnitude of the 1903 disaster occurred again until October 22, 1913.

Only two days before, on October 20th, the Stag Canyon Mine No. 2. had been inspected and the inspector reported that "the highest achievement in modern equipment and safety appliances that exists in the world" at Dawson and that the No. 2 Mine was "free from traces of gas and in splendid general condition." At about three in the afternoon on the 22nd, the homes of Dawson were rocked by an explosion that had occurred two miles away at the No. 2 Mine. Not only that, a 100-foot tongue of fire shot out of the mine's entrance the explosion was so intense.

Rescue efforts poured in from across the country to see how many of the 286 miners that went to work that day could be found alive. They found 23, leaving behind many widows and orphans. As it turned out, it would eventually be deemed the second greatest mining disaster of the century. Funerals for the miners went on for weeks in Dawson, and the cemetery had to be greatly expanded upon.

The town of Dawson also hosted its own unique version of La Llorona, the wailing woman of the Southwest. *Hispanic Legends from New Mexico* related a story of her on page 464: "One night two men were walking home from work. The morning dawn was just beginning to break through. Just as they approached the Catholic Church they saw a woman standing in front of it. Suddenly her cape flew open and they saw her body. It was a skeleton. Some people say it was la llorona coming back to get a priest to baptize her baby who was born a bastard and never baptized. She is supposed to appear there every so often until the priest will go with her and she will never rest until the baby has been blessed."

And yet, this tragedy didn't crush Dawson. The town actually continued to thrive, and peaked in 1918 with four million tons of coal. Tragedy struck again on February 8, 1923, and the Smikes had either left or were unable to help. This time, the accident occurred within the Stag Canyon Mine No. 1., when a mine train derailed and hit the supporting timbers of the tunnel mouth, which in turn ignited some coal dust. Of the 123 men in the mine, only two survived. Worse yet, a number of the men killed were the sons of the men killed ten years before. As such, their widowed mothers lost their sons as well.

Ultimately, the dead population was growing in Dawson while that of the living began to dwindle. Today, all that remains of Dawson is the cemetery, sporting 350 white crosses belonging to the dead miners. Although the town actually continued to thrive in spite of the disasters, the town shut down in 1950. Coal was no longer needed in the numbers that it once was, and there was no longer any need for poor Dawson.

Today, the ghosts of the miners can be seen frequently, and the Smikes, for all we know, retired with the town.

FATAL MINING ACCIDENT.

A Miner Blown to Fragments in the Zeile Mine.

A shocking accident occurred at the Zeile mine, at noon on the 24th of February, by which a miner named John Retallic was instantly killed. Deceased and several others were working in the 240-foot level of the mine. It is customary to explode blasts just before the dinner hour, so that the smoke will be cleared away by the time the miners are through with lunch. On the morning named, deceased and a fellow worker named Frank Spinetti, prepared four charges, and lighted the fuses, Retallic lighting three, and Spinetti setting fire to the fourth. They then retired to a safe distance and awaited the explosion. Two minutes elapsed, and three explosions were heard. Deceased then asked Spinetti if he had lighted his fuse; to which the latter replied, "Yes; but

I DON'T THINK IT HAS GONE OFF."

All the holes were charged with giant powder, but the fuse of one charge was somewhat longer than the others. Deceased had charge of the blasting operations, and impatient probably at the delay, went forward to ascertain the cause. Safety requires that at least five minutes should be allowed to elapse before venturing near an unexploded cartridge. This length of time, however, was not allowed to transpire in this case. Spinetti followed two or three paces behind Retallic, doubtless thinking that a due regard for safety demanded the exercise of a little caution. On reaching the fatal spot, Retallic remarked, "The

three charges have done first-rate." These were his last words. A few seconds later, and the dilatory charge went off, laden with death and destruction to the too venturesome miner. Spinetti was apparently on the edge of the explosion; or else, being in the rear of Retallic, he was thereby protected. He did not escape unhurt, however. He was knocked down by the concussion, or by the body of the deceased being blown against him, and crawled out on his hands and knees. He went to the station and gave the alarm. Owen Kelly, J. Dwyer, and others repaired at once to the scene of the accident. Retallic was found still breathing, but a

SHOCKING SIGHT.

The face was scarcely recognizable, the right eye was gone, and the right side of his head shattered to fragments. The left arm was blown out of all shape, the flesh hanging in shreds, and blackened with powder. The breast was mutilated in a terrible manner. The mangled remains were brought to the surface as quickly as possible, and afterwards taken to his boarding house, before reaching which, however, deceased breathed his last. Spinetti escaped with a few cuts about the head and legs, none of them being serious. He is able to be around, and will be ready for work again in a day or two. Retallic was a native of Devonshire, England, and about 32 years of age. He leaves a wife and three children in Cedar Rapids, Iowa. Has no relatives in this State. He had been in California about 18 months, and previous to his coming to this county was employed near San Jose. He had been working at the Zeile only three weeks. Work in levels was suspended for the remainder of the day out of respect to the deceased.

Amador Ledger-Dispatch,
2-26-1881, page 3.

16.
CALIFORNIA GOLD MINE HAUNTING
1881
JACKSON, CALIFORNIA

On December 9, 1881, the *Amador Ledger-Dispatch* of Jackson, California, ran a story about one of the town's gold mines being haunted by the ghostly spirit of a miner killed in the mine on February 24, 1881. At the time this happened, Jackson, California, was known for its hard rock mining, and the Kennedy Mine, which opened in 1860, was North America's deepest gold mine, eventually reaching a depth of 5,912 feet. Later in history, on August 27, 1922, Jackson experienced the worst gold mine disaster in United States history, when forty-seven miners were killed after they became trapped when a fire broke out in the Argonaut mine.

The site in which this incident happened was the Zeile Mine, also known as the Coney Mine or Zeila Mine, was located on the south side of Jackson, California, at the intersection of Broadway Street and French Bar Road. The mine began being worked in the 1860s and went on to produce more than $5 million in gold. According to *AmadorGold.net*, "By 1867, the mine had a 16-stamp mill and a chlorination plant. In 1890, the Zeile Mine was the main mine in Jackson, along with the Kennedy and Argonaut mines, which were a key part of the city's economic foundation. However, the Zeile Mine closed in 1914 due to a lack of a disposal area for tailings."

Historical marker at the Zeile Mine site. (Courtesy AmadorGold.net)

Interestingly, Jackson is located just four miles from the historic settlement of Sutter's Creek, which became the epicenter of California's Gold Rush, after discoveries were made in nearby Coloma in 1848. This entire area was honeycombed with valuable gold production.

But returning to the story of the haunting at the Zeile in 1882, local residents told of a mine worker, 32-year-old John Retallic, who died in a dynamite explosion inside the Zeile. The local newspaper, the *Amador Ledger-Dispatch*, ran this story about the accident on February 26, 1881: "FATAL MINING ACCIDENT. A Miner Blown to Fragments in the Zeile Mine. A shocking accident occurred at the Zeile mine, at noon on the 24th of February, by which a miner named John Retallic was instantly killed. Deceased and several others were working in the 240-foot level of the mine. It is customary to explode blasts just before the dinner hour, so that the smoke will be cleared away by the time the miners are through with lunch. On the morning named, deceased and a fellow worker named Frank Spinetti, prepared four charges, and lighted the fuses, Retallic lighting three, and Spinetti setting fire to the fourth. They then retired to a safe distance and awaited the explosion. Two minutes elapsed, and three explosions were heard. Deceased then asked Spinetti if he had lighted his fuse; to which the latter replied, "Yes; but I don't think it has gone off." All the holes were charged with giant powder, but the fuse of one charge was somewhat longer than the others. Deceased had charge of the blasting operations, and impatient probably at the delay, went forward to ascertain the cause. Safety requires that at least five minutes should be allowed to elapse before venturing near an unexploded cartridge. This length of time, however, was not allowed to transpire in this case. Spinetti followed two or three paces behind Retallic, doubtless thinking that a due regard for safety demanded the exercise of a little caution. On reaching the fatal spot, Retallic remarked, 'The three charges have done first-rate.' These were his last words. A few seconds later, and the dilatory charge went off, laden with death and destruction to the too venturesome miner. Spinetti was apparently on the edge of the explosion; or else, being in the rear of Retallic, he was thereby protected. He did not escape unhurt, however. He was knocked down by the concussion, or by the body of the deceased being blown against him, and crawled out on his hands and knees. He went to the station and gave the alarm. Owen Kelly, J. Dwyer, and others repaired at once to the scene of the accident. Retallic was found still breathing, but a shocking sight. The face was scarcely recognizable, the right eye was gone, and the right side of his head shattered to fragments. The left arm was blown out of all shape, the flesh hanging in shreds, and blackened with powder. The breast was mutilated in a terrible manner. The mangled remains were brought to the surface as quickly as possible, and afterwards taken to his boarding house, before reaching which, however, deceased breathed his last. Spinetti escaped with a few cuts about the head and legs, none of them being serious. He is able to be around, and will be ready for work again in a day or two. Retallic was a native of Devonshire, England, and about 32 years of age. He leaves a wife and three children in Cedar Rapids, Iowa. Has no relatives in this State. He had been in California about 18 months, and previous to his coming to this county was employed near San Jose. He had been working at the Zeile only three weeks. Work in levels was suspended for the remainder of the day out of respect to the deceased."

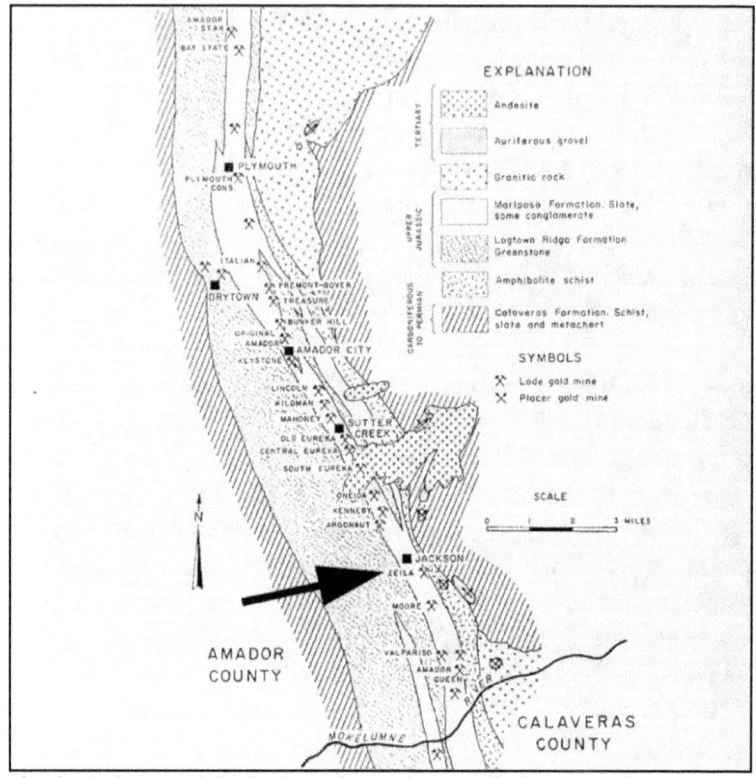

Geological map of the Jackson Area, showing Zeile Mine. (State of CA)

After his death, other miners reported hearing strange noises from the area of the mine where Retallic had been working on the day he died. Even though no work was being done by anyone in that area, the miners claimed to hear falling rocks, the rattling of drills, the clattering of mining picks, and some even claimed to hear groans.

Strangely, the eerie noises were never heard anywhere else in the extensive network of mine shafts. They were heard only in the exact spot where Retallic perished in the dynamite mishap that took his life. Also uncanny was the fact that the noises were only heard in the "dead hour of night" and were never heard in the daytime, although the mine was pitch dark at all of times of day and night.

Terror in the Mines!

Newspaper accounts said, "We are not prepared to vouch for the truth of this strange story, but it is said that a number of the miners will subscribe to its correctness from their own personal experiences. We think it not a very strange thing that ghosts should live about mines."

The Argonaut Gold Mine, located nearby. (State of CA)

A December 9, 1881 article in the *Amador Ledger-Dispatch* gave this summary of the mystery: "It is not generally known, even by the citizens of Jackson, that the Zeile mine, of this place, is subject to nightly visits from ghosts, spirits, or some other equally mysterious and strange visitants. But that such is the fact is vouched for by a number of miners who work in that mine on the night shift. It will be remembered that several months ago a miner by the name of [Retallic] was killed by a blast in this mine; and it is the exact spot where he met his sad and untimely death that is now infested with unknown and uncounted nocturnal occupants. While that portion of the mine is worked regularly during the day, and no unusual disturbances ever occur then to unsettle the nerves of the day workers, yet none of the miners are daring enough to remain there during the night hours. Those whose lot it is to work nearest this place allege that they hear emanating from that quarter, almost every night, falling rocks, rattling of drilling and clattering of picks, while some even claim to have heard groans, while it is known that no living persons are there, and no indications of disturbance can be detected by the miners who go to work there next morning. These phenomenal disturbances never occur except in this one particular spot, and then only in the dead hour of night, although it is as dark there during the day as it is at midnight. We are not prepared to vouch for the truth of this strange story, but it is said that a number of the miners will subscribe to its correctness, from their own personal experiences. If the affair is not a sensational hoax (and our informant seemed sincere in relating the matter to us), we would suggest that some of our courageous scientists get permission to go down into the mine and solve the mysterious problem."

Deaths in the Jackson area mines were frequent. Notices of mishaps among the mine workers appeared in the local papers almost weekly. Less than six months after Retallic's death, on August 6, 1881, another tragedy occurred in the Zeile, as reported in the *Amador Ledger-Dispatch*: "A young man named Frank Rogers perished in the Zeile mine on Saturday afternoon, between five and six o'clock. It was noticed that when the miners came up from their underground work, shortly before six o'clock, preparatory to the change of shifts, Rogers was not among the number, although it was known that he was in the mine. This was the first intimation that something was wrong. At the muster after the day's labor a comrade failed to put in an appearance, and the natural conclusion was that some disaster had befallen him. W. Boxall and N. Moon were lowered down the shaft, to explore for traces of the missing man. Soon their worst fears concerning the fate of the missing miner were confirmed. At one point in the shaft, they found Rogers' hat, and further down, at the 400-foot level, portions of brain were found. There was no room to doubt that he had fallen down the shaft, and that his body was in the sump, which contained over 39 feet of water. As speedily as possible work was commenced in grappling for the body; but more than three hours elapsed before they succeeded in recovering the remains. The body was brought to the surface about half past nine. Deceased fell, it is supposed, over 400 feet. How he came to fall must forever remain a mystery, as no one was near him at the time of the fatal mishap. There is, however, not the least reason to doubt that his death was purely accidental. The body was not so mutilated as might be supposed from a fall of such a distance. The top of the skull was crushed in, and the greater portion of the brain had escaped. One leg and one arm were also fractured; but the features were as placid and natural as could be. Deceased was only 20 years of age. His parents live at Grass Valley, Nevada county, and the remains were forwarded there on Monday. An inquest was held by the Coroner on Sunday, and a verdict of accidental death returned."

The deaths at the Zeile and its neighboring mines continued over the years, as shown in this newspaper article from the *Amador Ledger-Dispatch*, dated September 2, 1904: "The fatal accident at the Argonaut last week, causing the instant death of three men by a premature explosion, has naturally created some nervousness among some of the employees in sinking operations. The depression resulting from such a disaster will require some time to remove. It is now pretty well settled that the accident was not due to any defect in the fuse. The miners who are daily using the same fuse, and prefer it to all other brands on the market, do not accept the theory of a defective fuse. An examination of the shaft bottom after the debris of the fatal explosions was cleared away, revealed a condition of affairs that throws some light on the matter. All the holes on that side of the shaft where M. Quinn and W. Jewell were working were exploded. On the other side, where W. J. Curnow and A. Scatena had worked, only one hole exploded. That was the hole which they had had difficulty in loading, and which they had to resort to the blow-pipe to clean out. By this explosion a small portion of that side had been blown out, and it is supposed that the holes had mostly, if not all, been primed and lighted, but that the fuses of the unexploded holes had been blown out by the premature blast. The generally accepted explanation of the accident is either that this fuse had been accidentally lighted, or that in cleaning the hole the powder had been scattered around the mouth of the hole, and the hole had afterwards been insufficiently tamped or closed; that the lighted fuse communicated to the dry powder, which formed a train to the cartridge, and thereby exploded the hole almost as soon as it was lighted. The fuse has been found so uniform and reliable by experimentation before anil since the late catastrophe that it is regarded as improbable that a fuse eight feet long should be so utterly lacking in the ordinary characteristics of a primer as to cause the explosion."

Death could come in many ways, including explosions, falls, cave-ins, and falling rocks. Here is another account from the *Amador Ledger-Dispatch*, dated May 17, 1890: "George Casella, a miner employed at the Keystone, Amador City, was caved on and killed in that mine Thursday afternoon.... A large mass of rock fell, completely covering him, and when extricated, life was of course extinct. The body was mangled terribly, almost every bone being broken. The remains will lie interred today (Friday) in Sutter Creek."

As the mine shafts had to go down deeper and deeper, the cost of continuing the mining in the Jackson area skyrocketed, until economic factors began closing the mines. "The ground became extremely heavy at depth and required much timbering. As costs continued to increase during the early 1900s and were accelerated during World War I, a number of mines were shut down. The South Spring Hill mine was shut down in 1902, the Lincoln Cons., in 1912, the Oneida and Zeila [Ziele] in 1914, the South Eureka in 1917, and the Bunker Hill and Treasure in 1922," according to *The Gold Districts of California – Bulletin 193*, by William B. Clark.

With the closing of the Zeile in 1914 came an end to the legend of the ghostly miner that is said to have appeared there, beginning way back in 1881.

17.

TRAPPED IN A HAUNTED ARIZONA MINE

1893
NEAR TUCSON, ARIZONA

In March 1893, a number of major newspapers, including the *San Francisco Call*, reported on a strange event involving an abandoned Arizona mine that was supposedly haunted by the spirit of its murdered owner. The mine was named *Pais del Oro* (Spanish for "land of gold"), although no gold was ever found there, only low-grade copper. At the time this story first appeared, the mine - located near Tucson, Arizona - had been abandoned for many years, and all the local people living near it considered it to be "haunted." No amount of money would compel any of the locals to go down into the shaft.

According to the newspapers, "The mine is located in a most unexpected spot. It is not even near a mountain or hill, but on the floor of the Santa Cruz Valley, and a person could easily fall into some of the numerous shafts without seeing them. All windlasses and houses have been destroyed by lime and the raids of the Apaches. There is nothing to suggest that a mine exists except the opening. All the shafts except one are vertical and considerably over 100 feet deep. The other shaft is inclined and can be descended without much trouble."

The haunting of the mine is attributed to the grisly murder of its owner, some years prior to 1893. Since his name is not given in the reports, we will call the owner "Jones". According to the story, Jones had hired Native Americans to work the mine for him, and one of them became extremely upset at Jones. During an emotional confrontation, the miner smashed Jones' skull with a hammer. Not content with crushing his skull, the miner dragged the body onto a container of gunpowder. Reports said, "Placing a long fuse through the opening, [the killer] lighted it and made his own escape through a shaft. Another workman saw the whole affair and ran out another way, calling to all his companions who could hear to follow him. No living person was hurt, but the body of the dead man was blown to fragments."

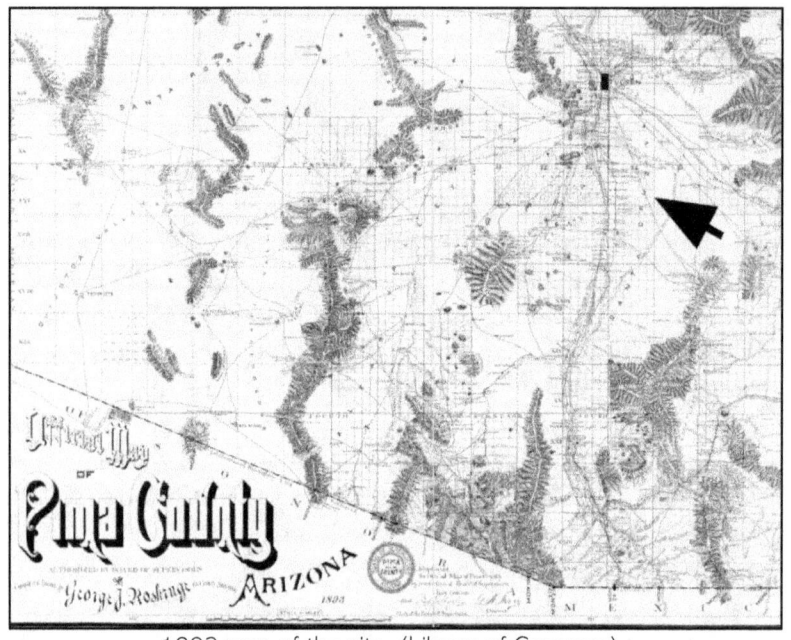

1893 map of the site. (Library of Congress)

Following the murder, local residents began to experience strange phenomena at the mining site, including seeing the image of the murdered Jones. "A few who have passed there at night say they saw the ghost of the dead man beckoning to them. He was not as he appeared in life, but in bleeding fragments, as he would be if he were put together after the explosion. It was a most horrible sight, and they also said they could hear the sound of the explosion every few minutes."

A LION WAS IN HIS PATH

AND FOR A TIME HIS CHANCES SEEMED EXCEEDINGLY SLIM.

A RANCHER'S ADVENTURE

He Enters a Deserted Shaft to Beat Up the Quarters of a Ghost and Gets Into Difficulties With a Mountain Lion Very Unexpectedly.

From the San Francisco Call.

"Pais del Oro" is the name of an abandoned mine in Pima county, Arizona. Where it got this name is one of the mysteries of that mysterious land, because it is not a gold mine, but a very poor copper one. However, that has nothing to do with the story. The mine is abandoned and has been for years, and you couldn't hire one of the Indians from the neighborhood to go down the shaft for any amount of money, or whisky even, because they say the mine is haunted by the ghost of its last owner.

That the mine is haunted is a generally accepted fact, as much so as the mine's name. The last owner was killed by an Indian workman who smashed his skull with a hammer. Not content with killing the man the murderer put the body on top of a can of powder, and placing a long fuse through the opening, lighted it and made his own escape through a shaft. Another workman saw the whole affair and ran out another way, calling to all his companions who could hear to follow him. No living person was hurt, but the body of the dead man was blown to fragments.

Mine entrance. (Library of Congress)

Especially eerie is the description of the spectral corpse. Given that the explosion essentially shredded Jones' body, the apparition is said to have looked like the various body parts, still bleeding, had been "put together after the explosion." The terror of the sight is well-described the persons that claimed to have seen it.

Following the murder, not a single person could be persuaded to go down into the mine shaft, either at night or in the day, for fear of encountering the animated corpse of the murdered mine owner.

Years passed, and a young Englishman named Charles Allen came to the area to engage in ranching. He soon heard the story of the haunted mine and grew interested in exploring the abandoned mine. Several times, he walked right up to the mine entrance, only to turn away.

Early in 1893, Allen was riding his horse past the mine entrance, feeling hot and exhausted from a long ride that had begun at daybreak and had lasted more than eight hours. He still had two more hours to go before arriving at his ranch and thought it would be a good idea to tie his horse to a cactus and go down into the shaft to cool off for thirty minutes or so. Although the story of the ghost was still in the back of his mind, he was less concerned because it was daytime. After some moments of thinking about it, Allen headed down into the mine shaft, wishing he could take his horse with him but was unable to do so.

"At the foot of the incline, about forty feet from the surface, a tunnel commenced, and as the air was cool and refreshing, Allen concluded that there was no use in going further. The interior of the mine was clean and dry and at the place where Allen stopped, there was plenty of light, although a few feet back in the tunnel it was dark, and gruesome bats flew past his head in an uncanny manner," according to the newspaper accounts.

Allen seated himself on the rocky floor, lighting his pipe and stretching himself out comfortably, watching the wreaths of pale blue smoke curl upward and pass out the top of the shaft. He felt pleased by his decision to take a few minutes to rest and was comforted by the fact that he could still see a small amount light filtering down from the mine entrance and also that the air was fresh and cool.

His thoughts turned to worrying about his poor horse that was still outside in the direct sunlight. Slowly, his thoughts became less and less distinct, and before he realized it, he was sound asleep. When he awoke, darkness had fallen, and the mine was totally dark.

"Allen did not open his eyes for a long time, and when he did, he did not know where he was. He was conscious that he was in a vault of inky blackness, the sides of which were made of cold stone. A feeling of horror came over him as He thought that he might have been buried alive."

Slowly, his mind cleared, and he remembered walking down into the mine and sitting down to rest. He remembered the stories about the ghost of Jones that appeared near this very spot, and he became uneasy. Reaching down for his revolvers, he remembered that he had left the pistols in his saddle holster on the horse.

He decided to head back up to the surface, but in the darkness, there was no clear indication of which way to go. He felt along the walls but could find nothing to guide him in the right direction. He considered that it might be best to wait in the mine until daylight.

As he felt along the wall, his hand found a piece of a candle that had been left behind by a previous visitor to the mine. He remembered that he had put matches in his pocket and was about to strike a match when he heard what sounded like an animal "purring" close to him. "He looked, and his blood almost froze at the sight of a pair of green eyes looking at him."

"He reeled and fell backward and the blackness around him turned red, while he could distinctly hear several loud detonations. Still the eyes glared, and soon he could see fragments of a human body forming underneath them. It was the ghost, and he felt that he would soon be insane."

Frightened and shaking, he managed to light a match and saw standing nearby an enormous mountain lion, looking very ferocious and hungry. In a moment, he realized that his mind had created the vision of the bloody ghost and the sounds of explosions. All of those imaginings vanished the moment he saw the lion.

"At the sight of something tangible, Allen's presence of mind returned, and he lighted the candle and took out a small pocketknife, determined to make as much of a fight as possible and not to die without letting the beast know whom he was killing. Allen set his candle down and looked at the lion and the lion looked at him. The beast was just at the bottom of the inclined shaft and blocked the way to the outside world. The lion did not seem in any hurry to make the attack, so Allen took up his candle and retreated into the tunnel in the hope that he could find one of the vertical shafts with enough ladder left for him to get out of the beast's way."

As Allen walked around, the lion followed him, keeping a distance of about ten feet from him. He suspected that the creature was scared of the light from his candle but feared that once the candle went out, he was lost.

"It must have been hours from the time the lion first appeared and the candle had burned down to only an inch or so.... He seemed to be passing the same place many times, but at last came to a pile of straw and a few feet ahead was a shaft with a plank across it. It was a ray of hope. He was sure he had not struck that place before and saw the means of escape. The candle was so short it almost burned his fingers, but he picked up a handful of the straw and started across the plank. It was old and nearly rotten, and beneath him were 100 feet of space. If the plank broke, it meant certain death."

THE LION LEAPED FOR THE END OF THE BOARD.

Tombstone Epitaph (May 23, 1897).

Although the wooden plank was old and rickety, Allen managed to get safely to the other side of the ravine and then pull the plank over to his side, leaving the lion to lash its tail in fury. The lion was obviously considering whether to try making the long jump over to where his prey now was standing, about ten feet away.

At this point, Allen had the idea to bait the creature to its death. He brought the wooden plank back over the hole, leaving it a few feet away from the lion, reachable with a jump. Then he took some of the straw he had picked up earlier, wrapped it around a rock, lit it on fire, and threw it over the ravine. It landed on the pile of straw beside the lion and set the straw ablaze.

"THE LION JUMPED FOR THE END OF THE PLANK."

Frightened, the mountain lion jumped and landed on the wooden plank, which Allen was steadying by standing on his end of it. When he saw that the lion had reached the plank, Allen hopped off his end of the plank, causing the board and the lion to go tumbling down into the 100-foot-deep hole. He could hear the dull thud of the heavy body and was certain that the fall had killed it.

Although the immediate danger was over, Allen was still in a fix, as he was trapped on the other side of the ravine, and the only tunnel within reach was a dead end. There was nothing to do but wait, which he did for nearly three days, at which time a search party organized by his friends finally found him. After finding his neatly starved horse standing near the mine's entrance, they deduced that Allen probably had gone inside.

After returning to his ranch, Allen rested for a couple of days. While recovering, he concluded that the mountain lion must have picked up his scent at the mine entrance and then followed him inside. Allen counted it a miracle that the creature had not attacked and eaten his horse, which was hitched to a cactus just a short distance away.

Was there any truth to the story? Arizona voting records establish the residence in Pima County, Arizona, of a person originally from England named Charles Allen, born in 1860. In 1893, he would have been 33 years old. Although we cannot be certain that this was the person named in the story, it seems probable.

CURIOUS LEGEND OF THE VIVARON SNAKE WORSHIP

Another Story of Human Sacrifice in Pecos Country

MYSTERIOUS CAVE SAID TO EXIST IN MOUNTAIN FASTNESS

"That snake story about the Zia pueblo reminds me of another Pueblo legend," said a well known citizen to the Journal yesterday. "It is called the legend of the Vivaron.

"I have not seen the legend in print and I doubt if it is generally known. It concerns the Glorieta Indians who formerly lived up in the Pecos country. Thirty miles north of Glorieta in a fastness of the mountains there is known to be a cave. This is not legendary, but a fact well known among residents of that section. This cave is walled up with a stone wall about fifteen feet thick. Through this wall there is a hole about six or eight inches in diameter.

The Butte Miner (July 14, 1906).

18.

GREAT MONTEZUMA'S GHOST!

1500s-1800s
PECOS PUEBLO, NEW MEXICO

From giant rattlesnakes to eternal flames, the legends of New Mexico's Pecos Pueblo are many. As if that wasn't enough, it was also rumored to be the birthplace of Montezuma. So the story goes, hundreds of years ago, a miraculous virgin birth produced the god-man known as Montezuma at Pecos. After performing various miracles, he departed the pueblo on the back of a giant eagle and flew south to found the Aztec Empire in Mexico. Before leaving, Montezuma lit a sacred, eternal flame. If it was kept burning by the people of Pecos Pueblo, one day he would return at dawn riding the same giant eagle. The flame was lit in a cave somewhere in the mountains, and to guard it was a rattlesnake of gigantic proportions.

While what was just recounted above was a folktale, historically speaking, Montezuma was born in Mexico around the year 1466. During his reign as emperor of the Aztec Kingdom, he famously encountered the Spanish conquistador Hernán Cortés. Despite an initially friendly if not cautious relationship, Montezuma's kingdom was conquered by Cortés and his men in the end. How Montezuma died is unknown, and some say he was killed by his own subjects for letting the kingdom fall to the Spanish.

Portrait of Montezuma, attributed to Antonio Rodriguez (1636-1691).

Terror in the Mines!

According to legend, shortly before the fall of his kingdom, Montezuma saw the proverbial writing on the wall. While he knew he could not stop the Spanish invasion, he could at least deprive them of the treasures and riches that they sought. In secret, Montezuma sent a procession northwards into what is today the American Southwest where a huge cache of gold, jewels, and other treasures were sealed away in a cave. The location of the cave has been speculated to be in many different places from Utah to Texas.

Railroad through Apache Canyon during the time of the Old West.

One legend said that Montezuma sent the treasure back to the place of his mythical birth, that of the area of Pecos Pueblo. Today, the spot of the alleged treasure cave might exist in an area known as Apache Canyon, near the famous Civil War battlefield of Glorietta. According to an article in *The Santa Fe New Mexican* of July 22, 1876, the treasure of Montezuma was valued at "a wealth of full two thousand millions of dollars" and comprised of "gold and silver, crown jewels and precious stones." The paper explained in fantastic fashion that,

Therefore [Montezuma] hastily dispatched the [treasure] in a trusty convoy secretly, by night, to be hidden in the mountains far to the north. It is related that the expedition journeyed for weeks upon its errand, and until it had reached a point about three hundred and seventy-five Spanish leagues to the north, whence it turned west and proceeded to the heart of the mountains, and buried the treasure in an immense mountain chasm there; and that the leaders of the expedition then put to death all the rank and file, fearing that upon their return they might by force or by treachery be made to discover to the avaricious conquerors the locus of the hidden wealth.

Sketch of the large Catholic Church constructed in Pecos after the conquest.

Naturally, the safest spot to store the treasure was in the same cavern as the monstrous rattlesnake and the eternal flame, faithfully watched over by the people of Pecos since Montezuma's departure. Actually, tales of the giant serpent were more prevalent than that of the treasure. Rumors circulated for many years that the people of Pecos regarded the giant snake as a god and even made human sacrifices to it. Notably, these stories surfaced after the desertion of Pecos Pueblo in 1838, when the last residents migrated to Jemez Pueblo to the west. So the story went, the eternal flame had gone out, and when that happened, there was no reason for the dwindling population of Pecos to remain there. And, when they left, they left the treasure where it sat still guarded by the fearsome snake.

Sketch of Pecos ruins by Lt. W.H. Emory.

A description of the snake's cave and a general location was given in *The Butte Miner* of July 14, 1906. An unnamed old-timer told the paper the following regarding the giant snake and the cave:

> "I have not seen the legend in print and I doubt if it is generally known. It concerns the Glorietta Indians who formerly lived up in the Pecos country. Thirty miles north of Glorietta in a fastness of the mountains there is known to be a cave. This is not legendary, but a fact well known among residents of that section. This cave is walled up with a stone wall about fifteen feet thick. Through this wall there is a hole about six or eight inches in diameter.
>
> "The legend says that many years ago when the Glorieta Indians were snake worshipers, a huge sacred serpent made his den back in this cave.
>
> "A human sacrifice was made periodically to this serpent, this ceremony being one of the most sacred tenets of the Indian religion. Sometimes the victim volunteered to sacrifice himself to this horrible death and sometimes he was placed on the altar by his fellow Indians."

Postcard depicting the remains of Pecos Pueblo, where Montezuma was once said to rule.

"You cannot get an Indian of any kind to go near that cave. As far as I know it has never been prospected. At least up to three years ago it had never been opened, to my knowledge," the old prospector concluded in reference to the alleged treasure.

The Santa Fe New Mexican of July 22, 1876, also noted the existence of the treasure cave, though they perhaps purposefully left out mention of the giant snake:

> And, come to think, we do recollect now once noticing a big hole or chasm in one of the sierras [mountains] hereabout, and not having seen this account of the cache though we had been wondering what became of those marvelous riches the Spaniards reported as existing at the ancient Mexican capital.

However, as with all legends of a fantastic nature, there are variations. As opposed to the historical Montezuma sending the treasure cache north to Pecos, there is a tale where the mythical Montezuma never left Pecos to begin with. In the case of that story, Montezuma became a treasure guardian, or what the Spanish called the *patrón*. This particular tale was unearthed by famous folklorist J. Frank Dobie, who heard it from José Vaca, a resident of Pecos in the early 20th century. According to Vaca in *Coronado's Children* on page 205, the *patrón* was "the dead man who guards the treasure. All these peoples long time ago who hide great treasure been careful to have *patrón.*"

Las Vegas, New Mexico, c.1900 at about the time the treasure hunt story told to Dobie would have taken place.

Vaca told Dobie of a Pecos Indian who had been jailed in Las Vegas on a rape charge. As the Pecos Indians' execution method for a rape crime was to strip the accused naked and tie them down over an ant den to be picked clean, the sheriff agreed to let the man get away in the night. Out of gratitude, the Pecos man told the sheriff of Montezuma's lost gold buried under Pecos Church.

In this variation, Montezuma was a half-Spanish half-Pueblo Indian chief who loved the Pueblo side of his lineage more so than the Spanish. Never mind that the historical Montezuma would have had no Spanish blood in him or live in New Mexico during the rule of the Spanish. And yet, somehow this Montezuma still possessed a great horde of gold like his Aztec namesake. This gold he wished to remain hidden until the Spanish had been driven from the land. The prisoner told the sheriff that the treasure was the "great secret of all the Indians of the Pecos Country."

137

The story went that when Montezuma was dying, he instructed his people to dig a big hole next to his pueblo and place his gold down in it. Next, the weakened old chief ordered his people to place him in the hole, so that his ghost could guard the treasure. Or in other words, Montezuma was buried alive with the gold to await death. His last words were that the gold was not to be unearthed until the Spanish had vacated the Indian's land.

Another postcard view of the ruins of Pecos Pueblo.

The prisoner told the sheriff that if he journeyed to the church at Pecos, off to the side of the road, he should find a distinct white rock. Within the rock he would find a small wooden cross wedged into one of the cracks. This signified the spot of the treasure, which was next to a black rock sitting over a cave. The instructions were to dig beneath the white rock, under which he would find the grave of Montezuma himself. Then, seven feet beneath Montezuma would be the treasure horde. However, the treasure was watched over not only by the great chief's ghost, but by secretive pueblo men as well. If the sheriff were to dig up the gold, he had best go at night with a car ready to speed away with the gold at once.

The sheriff rounded up his friend, José Vaca, and was ready to go dig for the treasure, but the sheriff's wife was too afraid of the treasure curse and begged her husband not to go. Vaca himself searched for the white rock on his own but could never find it, blaming Montezuma's ghost. "I cannot understand this *patrón*," he told Dobie on page 206. "I am strong man. [Montezuma] is dead, but he keep me off. I wish I know when he sleep."

Vaca never did find the treasure, but he did find a large white rock that used to have a wooden cross embedded in it in the area, so perhaps the story of Montezuma's grave from earlier had become conflated with this alleged treasure cave.

WHEW! .GRIM GHOST
FREQUENTS MINE AND
SCARES A MAN WHITE.

DOWNIEVILLE, Jan. 7.—What they stoutly claim is the ghost of S. W. Shaw, former owner of the mine, who was killed by an explosion, appears at the Rattlesnake property near this place with such frequency and grimness that miners and others will not go near the place.

Ed Deal decided the other day to stake out the claim formerly owned by Shaw. He took up his abode in Shaw's cabin and the first night was scared white by having his chair taken from under him by unseen hands, tables moved about and finally by seeing the mangled shape of Shaw appear before him in a threatening manner.

He swears the story is true and refuses to return to the spot.

Sacramento Californian (January 8, 1906).

19.

MINE HAUNTED BY DEAD OWNER

1906
DOWNIEVILLE, CALIFORNIA

The January 8, 1906 edition of the *Sacramento Californian* carried an article about a "grim ghost" seen frequently at the Rattlesnake Gold Mine, near Downieville, California, in Sierra County. The story said a number of miners had been terrified by the spectral apparition of a "mangled man." The ghost did not seem pleased to see them, and they felt threatened by it.

The witnesses claimed that the ghostly appearances happened "with such frequency and grimness that miners and others will not go near the place." One miner named Ed Deal, who did attempt to work the claim, soon found out why nobody else would step foot in it. Intent on spending the night in the mining cabin, he became alarmed when furniture in the cabin started moving around by itself, as if directly by invisible hands. When he attempted to sit on a chair, the chair was suddenly yanked out from under him.

Then the miner was shocked to see the ghost-like figure of a badly disfigured human appear before him. The figure seemed to be threatening him. Rushing out of the vicinity of the mine, Deal swore he would never go back there.

Many of the locals in Downieville believed that the ghost was that of one of their former residents who passed away tragically on May 29, 1906 at the very mine that he owned and where he worked. In life, he was known as Syvanno H. Shaw (S. H.

Shaw), originally from Vermont. A well-known prospector, Shaw was 71 years old and living in Downieville at the time of his death.

On June 8, 1906, Bert Dondero, an acquaintance of Shaw, went to Shaw's cabin to deliver some mail to him. Dondero entered the cabin and found Shaw dead, in a kneeling position beside his bed. His body was horribly mutilated, being disemboweled and having had one hand entirely blown off. Even after sustaining his horrific injuries, Shaw managed to drag himself up the mine shaft and into his cabin, where he died just before reaching his bed. It was in this position that his body was found. He had been dead for over a week before being found.

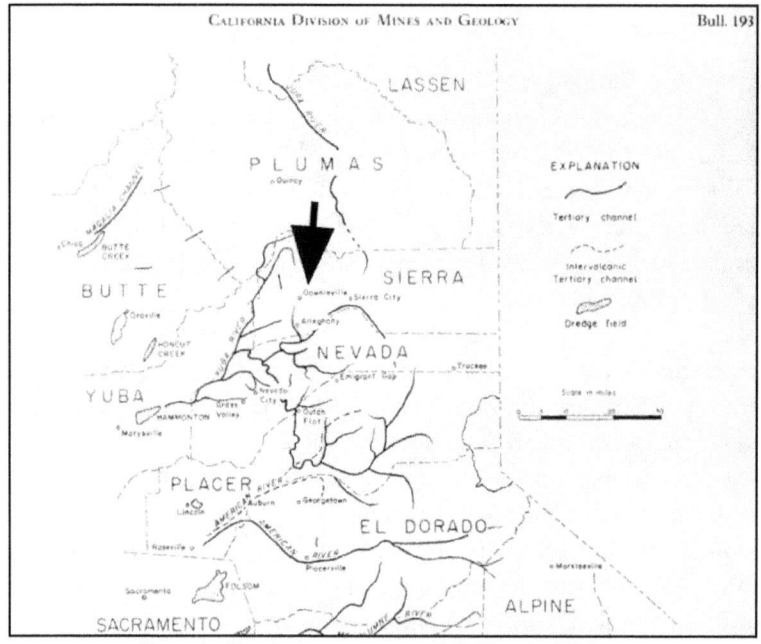

Downieville shown on map of California Mining.

The local coroner investigated the death and determined that Shaw had been killed by a powerful explosion that he himself had set in the shaft of the mine, about 140 feet down from where the body was found. The coroner deduced that Shaw had been setting off explosive powder as part of his mining, but one of his explosive charges had not fired. When he went to examine why it had not fired and to reload it, the powder went off.

A complete description of the death was given in the June 12, 1906 edition of the *Marysville (California) Daily Appeal*: "Terrible Death. Word was received last Friday afternoon that S. H. Shaw was found dead in his cabin at Rattlesnake that day. Bert Dondero, who is working at a claim about half a mile from the cabin of Mr. Shaw, went that day with some mail that had been brought up by Stanley Jenkins. When Mr. Dondero entered the cabin, he was startled on finding Mr. Slaw dead. He was in a kneeling position by his bedside Immediately, Mr. Dondero hastened to overtake Stanley to have him inform the coroner. The following day Coroner Archie McDonald, accompanied by August Costa and Otto Strand, went to the scene. On examination it was plainly seen that Mr. Shaw had been killed by an explosion of giant powder. After an examination of the body and surroundings in the cabin, the party then went to the tunnel, a short distance off. After going about 140 feet in the tunnel they discovered where the explosion had taken place, as there were evidences to indicate it. Portion of his clothing, besides his candlestick. etc., were found there. The body was in a terribly mutilated condition. One hand was entirely blown off and the lower part of the stomach was awfully mangled. The conclusion formed by the coroner and his assistants was that he had set off a round of holes, and had thought that one of them had not gone off, and he was on his way to reload it when in some manner the powder ignited, causing the explosion."

More details about Shaw's life were given, "The Coroner of Sierra County will leave Downieville this morning for Rattlesnake creek to take charge of the remains. He was an old prospector and had been working a claim which he owned for some time. He was well known and highly esteemed. His age was about 70 years. Besides being a member of the Grand Army of the Republic, he was a member of the Masonic fraternity, holding membership in Temple Lodge K. and A. M., of Sonora, and the funeral will undoubtedly be under the auspices of that order."

Following the tragedy of Shaw's death, several witnesses reported seeing his ghostly shape around the cabin where he died and near the entrance to the mine. The ghost's appearance was described as "mangled," and its behavior seems aggressive and threatening. Perhaps the spirit of Shaw resented other miners coming onto his former property and attempting to work the mine that he still considered his own.

The Downieville area, at that time, was one of the state's most productive gold mining areas. A report by the California mining commission said, "The most productive lode-gold districts in the northern end of the Sierra Nevada have been the Alleghany, Crescent Mills, Downieville, Forbestown, Graniteville, Grass Valley, Johnsville, Nevada City, and Sierra City districts."

20.

ROSWELL'S HAUNTED TREASURE HOLE

EARLY 20TH CENTURY
SOMEWHERE OUTSIDE OF ROSWELL, NEW MEXICO

Throughout the 1930s as part of the WPA, the Federal Writer's Project sent out reporters to interview the "old timers" as a way of both preserving our history and also garnering some entertainment. The tellers of the tales did not disappoint, and sometimes those FWP reporters dug up some wild tales, such as this one, which told of a cursed treasure cavern in Southeastern New Mexico. Interestingly, the nearest settlement was that of Roswell, today world-renowned as the site of a UFO crash in 1947. However, even Roswell with its rich Hispanic heritage has its share of folklore, from which this tale certainly descends. It was collected by Manuel Berg and given to him by a man identified only as Pacheco out of Albuquerque. Instead of paraphrasing, here is the tale in Pacheco's own words of a haunted treasure hole somewhere near Roswell:

"What I am about to tell you happened in the country around Roswell. There were three men, sheepherders, and one of the three was the foreman. They were in charge of a large flock of sheep and they very seldom came to town. At least they didn't ever get to town more than four or five times a year. This which I am telling was told by the foreman, and they all swore that it was the truth. Here is what the foreman said: "It was very late one night when I started to look for my two men, Pablo and Carlos. The stars weren't out and a strong wind had come up and I wanted to know whether the sheep were in a good shelter. You see, if they are not sheltered they become kind of wild and we have a lot of trouble gathering them together again. It took me a long time to find the two men because they had also read the signs that a storm was coming up and had herded all the sheep into a little valley.

Desert Scene
Southern New Mexico
As Seen From Highways No. 70 and 80

"When I did find the men they were trying to make a small fire but were having a little trouble. I helped them start the fire and put some coffee to heat. It was very cold and we weren't through with our coffee when it began to rain and blow real bad. We very quickly stamped the fire out and ran to find some better shelter for ourselves. We had already taken care of the sheep so we didn't bother about them. The three of us must have become separated because I soon heard an awful scream nearby and stopped to try and locate it. I called for Pablo and Carlos but only Pablo answered me and then came running over to see me. Carlos had gotten lost. We couldn't leave Carlos alone because one never knows what can happen to a person alone in the desert so we went back to where the fire had been and began to search. We searched very close to the ground, almost crawling on our hands and knees and even that way I almost fell into this hole. I was very scared when I found the hole but I called out if Carlos was there and he answered my call. Pablo came up to me and then we called to him again to find if there was anything we could do to help him. He, Carlos, said for us to tie a candle and some matches to a cord and lower it and at the same time to light a match so he could see where the entrance was. We did this and pretty soon we felt a jerk of the rope and knew he had found the candle.

Mysterious area outside of Roswell known as Lost River.
(Courtesy Historical Society for Southeast New Mexico.)

"Pretty soon a dim light came from this hole in the ground but we couldn't see anything from the top—I mean we couldn't make out anything that was down there. I called to Carlos that we would throw the end of a lariat down and then pull him up if he couldn't come up any other way. He said to wait a little bit because he had seen something which he wanted to investigate closer. Next thing he calls up to us and says that he has found a great number of sacks and boxes filled with money both gold and silver. While we were talking,

Carlos at the bottom of this hole and Pablo and I at the top, Carlos' light went out and we heard a strange voice say 'Don't be afraid. Many men have died for this treasure and many more are going to die'—and the voice began to laugh, a laugh so horrible that we almost fainted; that is, Pablo and I. Maybe Carlos did faint. Very much shaken I called to Carlos if he had the end of the lariat so we could pull him up. We didn't get an answer right away. Then we saw a light again and Carlos' voice said that he was coming up right away but first he would try to bring one of the sacks of money up with him. This time Carlos tied the lariat around his body and walked further into the cave. Then Carlos said that even one sack was so heavy that he couldn't lift it up. As he said this the candle went out again and the strange voice began to laugh, a laugh that makes my blood run cold and many times I wake up in the night and my heart is filled with great fear. After several moments the laughter died down and the voice said in a deep rumbling tone, 'You will never be able to take one sack. You must take the whole treasure at once or else nothing.'

"I know that Carlos fainted this time because we felt a strong pull at the lariat and frightened as we were we began to pull it up. It was a very hard job to pull Carlos up because he weighed almost two hundred pounds, and first we had to drag his body across the floor of the cave. By the time we got Carlos out, he had come to life again but he was a very sick man. He had a high fever and we got our horses and tied him onto his and came into town. Carlos never did get over his sickness. Within two weeks he was dead. Now Carlos being gone did not upset me a great deal because I thought it might have been a natural death but when Pablo became sick right after Carlos was buried, I became worried. I quit my job and left that part of the country but later I heard that Pablo had also died and his last words were that he had been cursed by an evil voice."

"Now," Mr. Pacheco said, "that is a story I heard from this man who lives in Martines town in Albuquerque.

This man also says that he has a map showing just where this treasure is hidden but I have never seen this map. When I asked him why he didn't go again to get the treasure, he said that since Pablo and Carlos were dead and he still lived, he preferred to go on living and let somebody else find the cursed treasure."

Truth be told, the strange story is reminiscent of many other folktales from New Mexico regarding buried treasure. If it has any truth or not is difficult to discern, but an interesting story it is nonetheless.

21.
PROSPECTOR ABDUCTED BY GHOST SHIP

MID 1878
INDIO, CALIFORNIA

G host ship sightings were a common staple of sailor lore, but they usually took place at sea, not in the skies, as happened in this case. This is also an especially intriguing story because it involves some of the well-known, tried-and-true themes of all good Western folklore, including prospectors, mining for treasure, a ghostly apparition, and the unforgiving desert.

The story appeared in the November 3, 1889 edition of the *San Francisco (California) Examiner*, on page 13, in an article titled "The Phantom Vessel. A Regular 'Flying Dutchman' Seen on the Colorado Desert." The story begins in Yuma, Arizona, late in the summer of 1878 when two German prospectors stumbled into town in a haggard and frenzied condition, excitedly proclaiming that a strange flying vessel had appeared over the desert and had kidnapped one of their comrades, taking him away into the clouds!

The *Examiner* reported, "Late in the summer of 1878, several years after the story of the discovery of the remains of a ship had been made public, two German prospectors reached Yuma from the Colorado desert. They were in a state of great distress and reported the loss of a companion on the desert. The prospectors, it seems, had been skirting the south and west sides of the San Bernardino range in search of minerals, and their companion was lost some six days before at a point about one hundred miles northwest from Yuma. The peculiar feature of their story was their associating with and attributing the disappearance of their comrade to an apparition which they had beheld the previous evening."

Gold prospectors c.1889. (Library of Congress)

The amazing narrative from the two exhausted prospectors continued in this manner: "About sundown, so the Germans said, and while encamped on the desert, they saw, at a short distance, an immense ship under full sail, which appeared to float before them as a cloud. She was of different form of construction from any vessel they had ever seen and was complicated and fantastic in her rigging."

In this part of the story, the witnesses describe what sounds like an ocean-going vessel ("an immense ship under full sail"), but they are careful to point out that the ship was of a totally "different form of construction" and that its rigging was "complicated and fantastic." Being of limited education and vocabulary, the witnesses were clearly struggling to understand and explain what they saw hovering in the sky above them.

SIGHTING THE "GHOST SHIP."

Clearly not a typical sailing ship, it was probably more akin to the mysterious airships that were sighted all over America in the late 1890s.

The *Examiner* article continued, "Their description of the vessel was by no means lucid, but they were very positive that their companion had been shanghaied and taken off on the 'ghost ship,' as they insisted on calling it."

This is the most fascinating part of the story – that the men believed their companion had been abducted by whatever entities were aboard the mysterious hovering airship. Although they did not give a detailed explanation of why they blamed the disappearance of their companion on the airship, it was clear that they felt the man had been "taken" ["shanghaied"] by the occupants of the bizarre ship.

The prospectors' strange story concluded as follows: "The story of the Germans was received with a good deal of contempt by the people of Yuma, who, after telling the prospectors that they were double adjective fools, sent two men and three Indian trailers on the train to Indio to search the desert east of that station for the missing man. The second day, his naked corpse was found about forty miles from the railroad, with the scorching rays of the sun falling upon it. He had died in the desert of thirst, but no sign of the phantom ship was seen."

Late in the summer of 1878, several years after the story of the discovery of the remains of a ship had been made public, two German prospectors reached Yuma from the Colorado desert. They were in a state of great distress and reported the loss of a companion on the desert. The prospectors, it seems, had been skirting the south and west sides of the San Bernardino range in search of minerals and their companion was lost some six days before at a point about 100 miles northwest from Yuma. The peculiar feature of their story was associating with and attributing the disappearance of their comrade to an apparition which they had beheld the previous evening. About sundown, so the Germans said, and while encamped on the desert, they saw, at a short distance, an immense ship under full sail, which appeared to float before them as a cloud. She was of different form of construction from any vessel they had ever seen, and was complicated and fantastic in her rigging. Their description of the vessel was by no means lucid, but they were very positive that their companion had been shanghaied and taken off on the "ghost ship," as they insisted on calling it.

The story of the Germans was received with a good deal of contempt by the people of Yuma, who, after telling the prospectors that they were double-adjective fools, sent two men and three Indian trailers on the train to Indio to search the desert east of that station for the missing man. The sec-

FINDING THE DEAD GERMAN.

ond day his naked corpse was found about forty miles from the railroad, with the scorching rays of the sun falling full upon it. He had died in the desert of thirst, but no sign of the phantom ship was seen.

The San Francisco (California) Examiner,
11-3-1889, p.13

Unbelievably, the missing German was found dead and without a stitch of clothing on him, leading some researchers to believe that he was abducted, possibly experimented on, and then dumped unceremoniously back onto the hot desert sand without clothing, food, or water, which was, in effect, a death sentence. In the UFO literature, there have been other cases where an abductee is returned to Earth sans clothing.

The story of the phantom airship and the abducted German prospector became quite the sensation in the North American newspapers, appearing in numerous papers in California, Nebraska, Kansas, Montana, North Dakota, and elsewhere. The earliest appearance of the article we could locate, however, was the previously mentioned *San Francisco Examiner* story of November 3, 1889.

22.

GHOST DOG OF KIEHL'S COAL MINE

SUMMER 1902
NORTH HAMPTON, OHIO

Our next tale took place in North Hampton Township, Ohio, in the summer of 1902. The newspaper article described "awesome happenings" comprising "Mysterious lights and strange animals." The primary witnesses used were the families of John Breitenstein and Peter Shaffer. In addition to those two families, it was stated that workers from Adam Kiehl's coal mine "could not be induced for love or money to enter the mine again" after seeing a ghost there. The paper made it clear that the area was an "improbable" location for a ghost, as Norton certainly was "not a deserted community with old houses and neglected-gardens, but on the contrary has fine farm houses with carefully kept lawns" and was "surrounded by rich, beautiful fields of grain."

The strange events began about ten years before the article was published, back when a man known only as Mr. Shaneman lived in small house across the road from the Breitensteins. Shaneman lived alone aside from his wife and between the two of them owned a nice farm. That year, Shaneman died and the farm was sold off. Being well-to-do, it was whispered that Shaneman was wealthy and had buried money somewhere on his property or had perhaps hidden it in his house. However, when the house was searched, only a measly $150 was found.

As opposed to being a treasure guardian to keep folks away, some thought the old man wanted his buried fortune to be found. The paper related, "Shaneman died very suddenly and there are those who say that the strange spectre about this place is the spirit of the old man trying to show where the money was hidden."

After this, the paper began the account of the supernatural happenings as told in the words of Mr. Bernstein, which shall be reprinted in full:

"We never saw any of these strange things before the death of Shaneman, and the first time I saw anything supernatural was the night after Shaneman's death. Peter Shaffer, John Mong and myself were sitting up with the corpse. Mong was smoking, and Shaffer and I had been talking. All of a sudden Shaffer gave me a little nudge and directed my gaze to the ceiling at the corner of the room where the corpse lay, when I saw a sight that fairly made my hair stand on end. What seemed a ball of fire had started from the corner of the room and was traveling slowly around the ceiling of the room.

"Did you see it?" said Shaffer. "Yes," said I. "Let's get out of here," were Shaffer's next words, and we made for home as fast as we could.

"And since that time the strange light has haunted this vicinity with the most unpleasant regularity. I have seen it many times, as has every member of my family and many others. It is more often seen in the winter than at any other time, but I have seen it twice this summer, the last time only a few weeks ago, when the thing looked into my bedroom window making the room as light as day, waking me up with the glare.

"We see this strange light at many times and places. Sometimes it rises out of the fields behind the Shaneman house, other times it rests upon the roof of the house. One night not long after the death of Shaneman, I, with my family, the family of Peter Shaffer, Shaffer, and other neighbors were sitting on the veranda of my home when suddenly a bright light as large as a street lamp rose out of the fields behind the Shaneman house, came up the lane, passed my home where we were sitting and went on up past my barn. Every one on the porch was silent, but as soon as the light disappeared, by common impulse, all were on their feet making for the barn behind which it had disappeared, but nothing could be seen.

"Then my son, Harry, who was married, moved into the house and he and his wife often saw the light. One night Harry came running over in breathless haste crying that our house was on fire. We rushed out to the back where a brilliant light was to be seen for a few moments, and then passed away into nothing.

"At one time we had an old apple tree which had blown down. One evening my son Milton came home late. He put up his horse and then came to the house. As he stepped into the door we saw that he was the color of death and whispered breathlessly, 'Oh, ma, come here!' My wife stepped to the door and there, playing about that old tree, were what seemed to be thousands of lights. They were about the size of candle flames and seemed to be of all colors of the rainbow. After a time they resolved themselves into balls of fire and rolled away down the orchard path.

"Peter Shaffer has told me of seeing the light many times and he isn't a man to lie," went on Mr. Breitenstein; "and this is the tale he told, me: Shaffer, with his wife and two daughters, were passing through the fields back of our house one night, bound for a neighbor's house. The women were walking ahead, when suddenly Shaffer saw the mysterious light moving along beside him. Then one of the girls looked back and with a scream started to run and soon the entire family was running for dear life over the fields. They came to a fence, but stopped not for a moment and how they ever got over the fence not one was ever able to tell. When they reached the road the light disappeared."

Ghost lights weren't the only supernatural happenings at the place, there was also a ghostly dog, possibly a hellhound. The witnesses who saw the ghost dog, Misses Louise and Minnie Shaffer, were returning home late one night when they saw "a strange animal walking along in front of them, creeping between their feet, plainly visible but of no substance, disappearing when they attempted to strike at it, but again walking with them a moment later. The girls took to their heels and arrived home almost dead with fright ad now no money could hire them to go out late at night alone."

MINERS AT WORK—TAMPING A CHARGE.

Mr. Shaffer also saw the ghostly canine and gave this account of it: "It was just before the abandonment of Adam Kiehl's coal mine, and some coal was still being taken from the mine. It was about 4 o'clock in the afternoon, when Shaffer saw someone going up the incline at the shoots, and as he wanted to get some coal he thought that would be a good time to go and order it. As he got near the mine, he saw the object hop up onto the platform on four feet. He thought, however, it was someone 'acting the fool,' as he expressed it, and went on. As he approached he saw the object was not a man, but some kind of a grey animal, which a moment later disappeared into the mine. He did not investigate, but made fast tracks for home."

A few days later, two men working in the mine saw the ghost dog and chased after it with their pics. "They would strike at it and it would fade into nothingness, but a moment later would be seen in another part of the mine," the article stated. "After a few fruitless attempts to approach it the men became frightened," and one of the miners "quit work and could not be induced to enter the mine again."

Other articles told of the animal, and were a bit more descriptive. Specifically, *The Newark Advocate* of August 11, 1902, said of it that, "It has the appearance of a wolf. On one occasion Geo. Conrad and Eugene Cady, of Barberton, pursued the animal but when they came near and struck at it the creature vanished."

That article stated that residents believed that the balls of light and the ghostly dog represented "the spirit of John Shaneman, and the theory is advanced by the superstitious that his shade has returned to earth to point out to relatives the hiding place of his fortune."

It reported, "John Breitenstein says the presence of the fireball recently awakened him at night and he saw it close beside his window. He rose and tried to follow it, but it vanished before going from his premises. Simultaneously with the seeing of the fireball the strange animal was seen, but in a different locality. It is said to haunt an abandoned mine near the Shaneman place. It has the appearance of a wolf. On one occasion Geo. Conrad and Eugene Cady, of Barberton, pursued the animal but when they came near and struck at it the creature vanished. Many people are visiting the locality to investigate the supposed apparition."

Unfortunately, what else they found, if anything went unreported.

23.
SILVER MINERS SEE UFO
NOVEMBER 1868
VIRGINIA CITY, NEVADA

Rather than terror in the mines, our next story could be considered terror above the mines. According to an article in *The Star and Enterprise* (Newville, PA.), on February 6, 1869, hundreds of silver miners, working the night shift, in Virginia City, Nevada, saw an incredible sight in the sky at 4 a.m. on November 27, 1868. The newspaper called the object a "celestial phenomenon." It was a star-like object, blood-red in color, with a bright, white halo surrounding it. From the lower part of the halo, a tail extended downward, looking like the curved blade of a saber. Witnesses said that both the halo and the tail were "filled with thousands of small and exceedingly brilliant stars."

The newspaper account said that atmospheric conditions were perfect for viewing the spectacle: "The sky was perfectly clear in every direction, and the eastern horizon was peculiarly blue and bright; not the slightest sign of mist above the ridges or the distant eastern ranges…."

The spherical part of the object, surrounded by the halo, appeared to be about five or six feet in diameter to an observer standing on the ground in Virginia City, while the tail was about eight feet long and about two feet in width at the top, tapering down toward a bluntly rounded lower end.

A CELESTIAL PHENOMENON.—The night gangs of men employed in the silver mines near Virginia City, Nevada, witnessed a beautiful celestial phenomenon at four o'clock on the morning of the 27th November. The sky was perfectly clear in every direction, and the eastern horizon was peculiarly blue and bright; not the slightest sign of mist about the ridges or the distant eastern ranges; yet the morning star rose of a blood-red color, and with a bright white halo, apparently five or six feet in diameter, surrounding it.— From the lower part of this halo extended downward a tail apparently eight feet long and two feet in breadth at the upper part. This tail was slightly curved, of a sabre shape, bluntly rounded at the lower end, and both it and the halo appeared to be filled with thousands of small and exceedingly brilliant stars. This strange light lasted some fifteen minutes, or until the star had risen so high that the tail of the halo appeared to be two or three feet above the crest of the distant range, when it suddenly faded out. The phenomenon was witnessed by hundreds of persons.

The Star and Enterprise (February 6, 1869).

The newspaper article said that "This strange light lasted some fifteen minutes, or until, the star had risen so high that the tail of the halo appeared to be two or three feet above the crest of the distant range, when it suddenly faded out. The phenomenon was witnessed by hundreds of persons."

Terror in the Mines!

What this strange object may have been remains a mystery to this day. It does not appear to have been a conventional astronomical event, such as a meteor shower or comet. No atmospheric disturbances were reported in any other regions of the country on this date. The sighting falls more easily into the realm of an unidentified flying object, similar to sightings that have been reported throughout America since the 1940s.

A Nevada silver mine. (Public Domain)

There was, however, a meteor shower reported two weeks earlier in the neighboring town of Gold Hill, Nevada, located just two miles south of Virginia City. The *Daily News* of Gold Hill reported that the meteor shower occurred on November 14, 1868, saying, "This morning between 12 and 4 o'clock, the grandest meteoric shower ever seen in this State occurred; and all who witnessed it express the greatest admiration of its magnificence."

A large part of onlookers gathered at Gold Hill to watch the meteors, which "kept up their nocturnal illuminations for four hours, and hundreds of them could be seen, comet-like, flying through endless space."

The November 14 meteor shower was also witnessed in San Francisco, California, about 250 miles to the southwest, according to the *Daily News*. "An extraordinary display of meteors was visible last night, lasting from an early hour in the evening until nearly daylight. The heavens were aglow with shooting stars, leaving trains of light sometimes visible for several seconds. The general course was from east to west. At about 1 o'clock a brilliant meteor burst, leaving a thick cloud, which slowly vanished from view."

Could the strange object sighted two weeks later in Virginia City have somehow been related to meteor activity? This remains uncertain.

24.

HEADLESS HORSEMAN
OF THE KIAMICHI MOUNTAINS
1860s-1890s
KIAMICHI MOUNTAINS, OKLAHOMA

One of the more unique treasure guardians of the Old West was a headless horseman of the Choctaw Nation. The headless rider served as an omen to anyone who might try to disturb a treasure of minted U.S. gold coins. The treasure had a complicated history in that it was part of a $2 million settlement with the Choctaw Nation on behalf of the U.S. government, who had seized unallotted lands from them. The payment was made in 1858, with $250,000 worth of it being issued in gold coins. The rest was to be given in the form of bonds over the next several years. However, when the Choctaw sided with the Confederacy during the Civil War, the deal was rendered moot. Thus, all they had was the cache of gold coins which they buried.

The gold was buried somewhere near the Kiamichi River in Oklahoma under the orders of George Hudson, principal chief of the Choctaw Nation at the time. He ordered the gold to be hidden because the Union Army was on the move, and he feared they might seize the gold coins. To do so, Hudson sent out two men of the tribe to secret the gold away, which they did by hiding it in a small cave near the river. Over the course of the war, several of the men who had hid the coins died in the fighting, leaving Escar Colbert as the lone survivor by the time the war ended.

George Hudson, principal chief of the Choctaw Nation.

Though a map to the treasure had been made by the Choctaw Council, during several swift and forced moves, it had been misplaced over the years. When Colbert returned from the war, badly mangled, he was unable to lead anyone back to the cave itself, and could only offer vague directions from a memory now clouded by the trauma of war. Worse yet, word of the treasure had spread, and the Kiamichi Mountains were already being scoured by outlaws on the hunt for it.

Colbert, having regained some strength, led an expedition in search of the cave again in 1867 with several other Choctaw men. After several days of scouring various caves, they finally found the right one... or so it seemed. In the dead of night, they descended the mountain on which the cave resided, carrying with them what appeared to be bags of gold. Unbeknownst to them, outlaws were watching. The men were former Quantrill raiders from Missouri. With their rifles at the ready, they began a swift and deadly massacre on the mountain. Colbert was atop his horse when he was killed. An outlaw popped out of the brush with a double barreled shotgun aimed at Colbert's head. It hit his neck, blowing his head right off his body. The spooked horse then took off with the headless corpse still attached.

The former guerillas massacred every last one of the Choctaw men, but they were in for several surprises. The sacks of gold were empty. As it turned out, Colbert and his men had entered the wrong cave and came out empty handed. In the darkness, it only appeared that the sacks were full of loot. The outlaws soon regretted their actions for another reason. In the nearby brush, they began to hear hoofbeats. Then, suddenly, only yards away, the horse with the headless rider dashed past them. The dead body should have fallen off the horse long ago. Not only that, to them, it looked as though it was actually in command of the angry steed. The outlaws fled the area and never returned, though they made sure to tell the tale of their encounter to their friends.

Eventually, tales of the massacre reached the leaders of the Choctaw nation. Knowing the general area of the attack, they went to investigate, both to bury the dead and hopefully the gold. They found three of the men, but no gold. Nor did they find Colbert's severed head. Or his horse. Or his headless corpse. Soon after, anyone who passed through the mountains was liable to see Colbert's angry, headless ghost race down the mountain on his horse. Until he was avenged, he was doomed to ride the mountain at night forever the Choctaw believed. He was also destined to guard the treasure.

Over the years, several men braved the mountains to look for the lost gold. Among them were James Calhoun Meador, who hunted for the treasure under the endorsement of the Choctaw Nation, the agreement being that Meador could keep twenty percent if he found it. He never did, but he did see the headless horseman ride down the mountain on four different occasions.

After Meador gave up, next came James Barnett, who unlike his predecessors, hit paydirt in October of 1893. While digging in a cave in the mountains, his shovel struck something hard. Barnett soon unearthed twenty $50 gold coins minted in the year 1858. This was indeed the lost treasure, which Barnett knew of being one-fourth Choctaw himself. The only problem was that Barnett was by then utterly exhausted. On the trail of the treasure, he had gone without food for three days and couldn't even think of excavating the whole cache. As such, he stored what gold coins that he could in the pockets of his overalls and stumbled down the mountain back to his camp. By sundown the next day, Barnett mustered enough strength to mount his horse and ride down the mountain. As he did, he heard ominous hoofbeats. He turned to see a headless rider storming by in the dusk, with just enough light present to make out the man's bloodstained shirt.

Barnett rode like mad for Tuskahoma and decided to leave the rest of the treasure where it laid. Two years later, Barnett had a change of heart. He was itching to get back to the gold, maybe the headless rider had finally given up the ghost. Though it took three more years, Barnett eventually found his way back to the old cave in 1898. No sooner than he had refound the cave, there came the headless horseman dashing towards him again. That time Barnett left and didn't come back for good.

Barnett was lucky compared to other men. Mysterious deaths seemed to plague the region, with men found dead as though they had been scared to death. Others committed suicide, it appeared. Was it the headless rider at it still?

The tale only survived into the 20th century thanks to Maurice Kildare, who heard it from his grandfather, James Calhoun Meador. When Kildare asked his grandfather about the headless horseman for the last time, Meador simply replied, "Them mountains is filled with evil spirits."

25.

UNDEAD ARMY OF SHIPROCK

1600s-1890s
DINÉTAH

Amidst the many landmarks dotting Dinétah, the homeland of the Navajo, Shiprock Peak stands among the tallest, literally and figuratively. Stretching just a little over 1,500 feet into the sky, the formation was so named due to its sailing-ship-like appearance in the distance. In Navajo it is called *Tsé Bit'a'í*, literally meaning *rock with wings*. That name is appropriate, for according to Navajo legend, the peak didn't always reside in New Mexico's San Juan County. The formation was called the "rock with wings" because, according to myth, it literally flew to the spot to transport the Navajo to a new home. Some have likened this story to the "ancient aliens" class of thinking, with the idea being that a spacecraft of some kind transported the Navajo. Others view it simply as a myth, and instead of a craft, Shiprock peak was the remains of a mighty bird monster that had carried the tribe to New Mexico.

Whatever Shiprock originally was, in the early days, the Navajo lived atop the peak for safety, only venturing below to gather food. But then, one day, calamity struck in the form of a horrible storm. An intense lightning strike obliterated the bridge allowing access to the peak, and in the process, hundreds of Navajo women and children were left stranded. Lest they disturb their *chindi*, or ghosts, the Navajo today consider the peak sacred and unscalable.

WHERE MAN HAS NEVER TROD.

Government Expedition in Search of a Modern El Dorado.

A TOMB THE NAVAJOS HAVE GUARDED WITH THEIR LIVES.

A Fabled Land of Hills and Mountains of Solid Silver and Gold—Whence No White Prospector Has Ever Returned Alive — The Navajo Legend of the Grave of the Indian "Mother of Life."

FORT WINGATE (N. M.), April 19.—The Mother of Life is to be disturbed. Her tomb is to be profaned by the insulting presence of the white man.

If the white men like the tomb the indications are that they will take it, for the evil spirit of trade has entered the souls of her Navajo warriors, and though they have guarded the tomb of the Mother of Life with their bodies ever since the morning of time, the day seems to be coming when the holy mountains will be traded for fields and pasture lands.

An expedition is now preparing to go from

San Francisco Examiner (May 1, 1892).

Terror in the Mines!

A gold strike in the nearby Carrizo Mountains in the year 1890 caused a particularly interesting Navajo legend regarding Shiprock to resurface. In summary, during a blizzard, a group of prospectors had sought out a lost gold mine somewhere on the Navajo Reservation and actually found it on Carrizo Mountain in March of 1890. The Navajo, led by Chief Black Horse, tried and failed to expel the miners. As such, Chief Black Horse traveled to Fort Wingate to get help from the soldiers that oversaw the reservation. Troops were sent out to arrest the miners, who brought back to civilization proof of their gold-strike. When the news hit, it caused considerable excitement, so much so that there was talk of relocating the Navajo so that the reservation could be thoroughly mined.

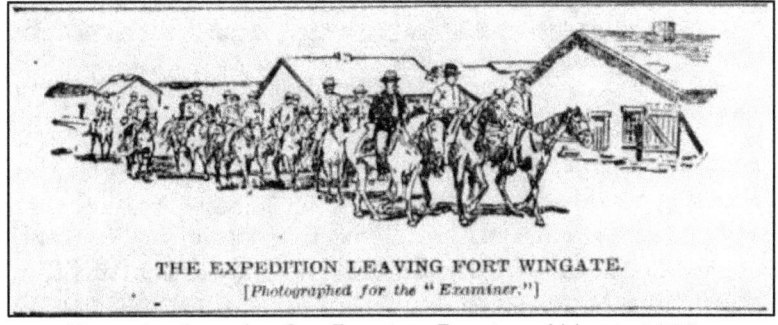

THE EXPEDITION LEAVING FORT WINGATE.
[*Photographed for the "Examiner."*]

Illustration from the *San Francisco Examiner* of May 1, 1892.

The *San Francisco Examiner* of May 1, 1892, reported on a government sanctioned expedition from Fort Wingate which explored the Carrizo Mountains and brought back further ore samples of valuable metals. Politicians began talking about dividing up the Navajo reservation so that it could be mined, which naturally upset the Navajo. Worse yet, if the white man found gold on Carrizo Mountain, could the mining of Shiprock Peak be next?

Thankfully, this never came to pass. The Navajo kept their sacred lands, and Shiprock Peak went unmolested. As stated before, the scare brought to light an old Navajo legend dating back to the days of Spanish rule and the Conquistadors. Supposedly, Spanish priests had enslaved local Navajo to work a gold mine very near Shiprock in the 16th century.

Shiprock Peak with Navajo in foreground. (Library of Congress)

The *San Francisco Examiner* of May 1, 1892, reported the legend. For context, the *Examiner* referred to Shiprock as "The Tomb of the Mother," and treated the mountain as though it was one of the chief Navajo deities that would one day cause an earthquake to drive away invaders from the sacred Navajo lands. The article related:

> There is a tradition that for fifty years the holy ground was not in the possession of the Indians. That must have been about two centuries ago. There is a lost Padre mine somewhere in those mountains. The old Spanish Fathers are said to have worked it for generations. According to the story there was a strong army at the mine against which the Navajos could do nothing, for the men wore iron breastplates that their arrows could not pierce. During all this period the mother was restive. For weeks the sun did not come.
>
> Ship Rock mountain and the surrounding country heaved and shook with the mother's indignation. Mountains opened and rivers ceased to run, but the invaders stayed on. At last the turbulence awoke the Thunder God and the God that-shakes-the-earth. The

mother invoked their aid, and then came a day when the sky opened and let down a deluge, and the mine was crushed between two mountains. Then all the Navajos who had been killed in fighting the invaders were awakened from death by the shaking, so fierce it was, and they fell upon the armored soldiers and drove them into a canyon, from which they could not escape, and rolled down bowlders on them until they were all dead and buried.

Ever since that time the Navajos have been in possession of the Ship Rock peak.

Depiction of forced Native American slave labor.

It's unknown if the *San Francisco Examiner* invented this tale of undead warriors for the delight of readers, or if it was a real oral tradition among the Navajo. Regardless of where the story came, it's certainly a fantastic one.

BIBLIOGRAPHY

Books

Anam Paranormal. "114 - Lafeyette Cemetery - The Vampire Grave." https://www.anamparanormal.com/114

Clark, Jerome. *Unnatural Phenomena: A Guide to the Bizarre Wonders of North America.* Santa Barbara, CA: ABC-CLIO, 2005.

Dobie, Frank J. *Coronado's Children: Tales of Lost Mines and Buried Treasure in the Southwest.* University of Texas Press, 1978.

-----------------*Apache Gold & Yaqui Silver.* Little, Brown and Company, 1939.

Gregg, Josiah. *Commerce of the Prairies.* University of Oklahoma Press, 1958.

Lewis, Chad. *The Van Meter Visitor: A True and Mysterious Encounter with the Unknown.* On The Road Publications, 2013.

Robe, Stanley L. *Hispanic Legends from New Mexico (Folklore and Mythology Studies: 31).* University of California Press, 1980.

Valley-Fox, Anne and Ann Lacy (Compilers). *Lost Treasures & Old Mines: A New Mexico Federal Writers' Project Book.* Sunstone Press, 2011.

Articles

Kildare, Maurice. "Kiamichi Warrior's Gold." *Frontier Times* (December-January 1972).

Kilen, Mike. "Van Meter remembers 1903 visit from winged monster." *Des Moines Register.* (July 2015) www.desmoinesregister.com/story/news/2015/07/01/van-meter-remembers-1903-visit from-winged-monster/29583469/

Videos

KUSA Staff. "There's a vampire buried in Lafayette?" (KUSA, 2015) https://www.9news.com/article/news/theres-a-vampire-buried-in-lafayette/134429175

Mr. Mythos. "Treasure Hunting with Demons." YouTube (July 9, 2022).

INDEX

ABOUT NOE TORRES

Noe Torres is a recognized expert in the field of UFOs and the paranormal. He is an author, publisher, and member of the Mutual UFO Network (MUFON). He holds a Bachelor's in English and a Master's in Library Science from the University of Texas at Austin. He has written one of the most popular books about the famous Roswell Incident, titled *Ultimate Guide to the Roswell UFO Crash*, which is the top selling book among tourists visiting Roswell, New Mexico. He has also written several other well-reviewed books, including *Mexico's Roswell*, *The Other Roswell*, *Aliens in the Forest*, *Fallen Angel*, and *The Coyame Incident*.

Noe has appeared on several nationally-broadcast television shows, including season 2, episode 1 of the Travel Channel's *Mysteries of the Outdoors*, titled "Strange Attraction," which premiered in August 2017. In that show, he is interviewed extensively about unexplained mysteries in Big Bend National Park. Also, in 2017, Noe was featured in an episode titled "The Marfa Lights" for the TV series *Mysteries of the Unexplained*. In 2008, he appeared in season 1, episode 4 of the History Channel's *UFO Hunters*, in a show called "Crash and Retrieval."

Noe has appeared several times on George Noory's famous radio show *Coast to Coast AM*, as well as on The Jeff Rense Program and may other shows. He is also in high demand as a speaker at UFO and paranormal conferences and festivals, having been a featured speaker at the 2017 International UFO Congress in Scottsdale, Arizona. He has also spoken five times at the annual Roswell UFO Conference and at many other UFO conferences throughout the United States and Mexico.

ABOUT JOHN LEMAY

John LeMay was born and raised in Roswell, NM, the "UFO Capital of the World." He is the author of over 50 books, many of them on the history of the Southwest such as *Tall Tales and Half Truths of Billy the Kid*, and *Roswell USA: Towns That Celebrate UFOs, Lake Monsters, Bigfoot and Other Weirdness*. In addition to non-fiction, he is also the author of the novels *The Noted Desperado Pancho Dumez* and *Once Upon a Time in Fort Sumner*. He is also the editor/publisher of *Strange West Magazine* and has written for Western journals and magazines such as *True West*, *The Coalition Journal*, the *Tombstone Epitaph*, and the *Wild West History Association Journal*. He is a Past President of the Board of Directors for the Historical Society for Southeast New Mexico.

The following titles are available for purchase on Amazon.com, and are available to bookstores at a wholesale discount via Ingram Content Group (ISBNs of available editions listed for this purpose)

CRYPTOZOOLOGY/COWBOYS & SAURIANS

Cowboys & Saurians: Prehistoric Beasts as Seen by the Pioneers explores dinosaur sightings from the pioneer period via real newspaper reports from the time. Well-known cases like the Tombstone Thunderbird are covered along with more obscure cases like the Crosswicks Monster and more. Softcover (357 pp/5.06" X 7.8") Suggested Retail: $19.95 ISBN: 978-1-7341546-1-0

Cowboys & Saurians: Ice Age zeroes in on snowbound saurians like the Cerato-saurus of the Arctic Circle and a Tyrannosaurus of the Tundra, as well as sightings of Ice Age megafauna like mammoths, glyptodonts, Sarkastodons and Saber-toothed tigers. Tales of a land that time forgot in the Arctic are also covered. Softcover (264 pp/5.06" X 7.8") Suggested Retail: $14.99 ISBN: 978-1-7341546-7-2

Southerners & Saurians takes the series formula of exploring newspaper accounts of monsters in the pioneer period with an eye to the Old South. In addition to dinosaurs are covered Lizardmen, Frogmen, giant leeches and mosquitoes, and the Dingocroc, which might be an alien rather than a prehistoric survivor. Softcover (202 pp/5.06" X 7.8") Suggested Retail: $13.99 ISBN: 978-1-7344730-4-9

Cowboys & Saurians South of the Border explores the saurians of Central and South America, like the Patagonian Plesiosaurus that was really an Iemisch, plus tales of the Neo-Mylodon, a menacing monster from underground called the Minhocao, Glyptodonts, and even Bolivia's three-headed dinosaur! Softcover (412 pp/5.06"X7.8") Suggested Retail: $17.95 ISBN: 978-1-953221-73-5

UFOLOGY/THE REAL COWBOYS & ALIENS IN CONJUNCTION WITH ROSWELL BOOKS

The Real Cowboys and Aliens: Early American UFOs explores UFO sightings in the USA between the years 1800-1864. Stories of encounters sometimes involved famous figures in U.S. history such as Lewis and Clark, and Thomas Jefferson. Hardcover (242pp/6" X 9") Softcover (262 pp/5.06" X 7.8") Suggested Retail: $24.99 (hc)/$15.95(sc) ISBN: 978-1-7341546-8-9\(hc)/978-1-7344730-8-7(sc)

The second entry in the series, *Old West UFOs*, covers reports spanning the years 1865-1895. Includes tales of Men in Black, Reptilians, Spring-Heeled Jack, Sasquatch from space, and other alien beings, in addition to the UFOs and airships. Hardcover (276 pp/6" X 9") Softcover (308 pp/5.06" X 7.8") Suggested Retail: $29.95 (hc)/$17.95(sc) ISBN: 978-1-7344730-0-1 (hc)/ 978-1-7344730-2-5 (sc)

The third entry in the series, *The Coming of the Airships*, encompasses a short time frame with an incredibly high concentration of airship sightings between 1896-1899. The famous Aurora, Texas, UFO crash of 1897 is covered in depth along with many others. Hardcover (196 pp/6" X 9") Softcover (222 pp/5.06" X 7.8") Retail: $24.99 (hc)/$15.95(sc) ISBN: 978-1-7347816-1-8 (hc)/978-1-7347816-0-1(sc)

Featuring cases the authors missed, *The Lost Cases* covers things such as the skyquakes recorded by Lewis and Clark, airships and the Spanish American War, Pancho Villa and crystal skulls, lost alien tribe of the Tundra, invisible alien monsters, the Great Moon Hoax of 1835, hellhounds and airships, the Sonora Airship Club and more. Softcover (252 pp/5.06" X 7.8") Suggested Retail: $18.99 ISBN: 978-1-953221-55-1

COWBOYS & SAURIANS CONT'D

Cowboys & Saurians: Dinosaurs Down Under takes the series to Australia to explore tales of the cattle devouring Burrunjor, the dreaded Diprotodon, the terrible Tantanoola Tiger, the marsupial Sasquatch known as the Yowie, plus Thylacines, Bunyips, giant rabbits, Megalodons and dinosaurs in nearby New Zealand. Softcover (240 pp/ 5.06" X 7.8") Suggested Retail: $14.95 ISBN: 978-1-953221-34-6

As the title suggest, *Cowboys & Saurians in the Modern Era* takes the series into the 20th Century with tales of the Texas Pterosaur flap of 1976, the Bladenboro Beast of the 1950s, the Busco Turtle Beast of the 1940s, dinosaur sightings in the Great Depression and far out tales of mini-mastodons, dinosaur men, and Snallygasters. Softcover (320 pp/ 5.06" X 7.8") Suggested Retail: $19.95 ISBN: 978-1-953221-22-3

Settlers & Serpents wrangles the best "Snaik Stories" of the Southwest and beyond in a single volume. Whether it's simple giant snakes or lake serpents, they're corralled in the pages within. Also included are entries on the Leviathan in Mesoamerica and the Southwest plus a detailed look at the giant rattlesnake of Pecos Pueblo. Softcover (180 pp/ 5.06" X 7.8") Suggested Retail: $14.99 ISBN: 978-1-953221-21-6

Written for young readers ages 9-12, *Monsters of the Old South* collects the best creature stories of the swamplands including the White River Monster, Green Eyes, the Crocodingo, the Averasboro Gallinipper, the Tennessee Snake Woman, the Arkansas Gowrow, Bigfoot in the Mississippi River and more. Softcover (122 pp/4.25" X 7") Suggested Retail: $12.99 ISBN: 978-17347816-9-4

THE REAL COWBOYS & ALIENS CONT'D

Early 20th Century UFOs kicks off a new series that investigates UFO sightings of the early 1900s. Includes tales of UFOs sighted over the *Titanic* as it sunk, Nikola Tesla receiving messages from the stars, an alien being found encased in ice, and a possible virus from outer space!Hardcover (196 pp/6" X 9") Softcover (222 pp/5.06" X 7.8") Suggested Retail: $27.99 (hc)/$16.95(sc) ISBN: 978-1-7347816-1-8 (hc)/978-1-73478 16-0-1(sc)

UFOs in the Roaring Twenties takes a look at UFO sightings in the 1920s just as the title suggests, along with accounts of Mothman in Nebraska, Lincoln LaPaz's first UFO case, Men in Black investigating an airship crash in Braxton County, West Virginia, Camden's Cosmic Sniper, and much more! Softcover (248 pp/5.06" X 7.8") Suggested Retail: $19.99 ISBN: 978-1-953221-51-3

UFOs of the Turbulent Thirties concludes the authors' investigation of the last unexplored decade of Ufology in the Great Depression with accounts of Mothman, Ghost Fliers, Nazi Bells, the Underground City of the Lizard People, a vanished village on the tundra, and even gangsters and aliens. Softcover (212 pp/5.06" X 7.8") Suggested Retail: $17.95 ISBN: 978-1-953221-35-3

Written for young readers ages 9-12, *Space Monsters of the Old West* collects the best alien sightings of the Wild West including Mummies from Mars, Bigfoot from the Moon, Pascagoula's space ghouls, the Crawfordsville Monster, Spring-Heeled Jack, Blobs from space, and even the dinosaurian alien creatures that invaded Van Meter, Iowa. Softcover (120 pp/4.25" X 7") Suggested Retail: $12.99 ISBN: 978-1-953221-87-2

COWBOYS & MONSTERS

Cowboys & Monsters features potentially true stories of real vampires, werewolves, and even mummies unique to America's Wild West period. Examples include the cursed mummy of John Wilkes Booth, New Orleans immortal vampire Jacques St. Germain, precursors to the Beast of Bray Road, and the origins of Skinwalker Ranch. Softcover (316 pp/5.06" X 7.8") Suggested Retail: $19.99 ISBN: 978-1-953221-46-9

The first entry in this trilogy of non-fiction terror sinks its teeth into the lore of the vampire in North America and Mexico, with detailed rundowns on the vampire hunters of Exeter, Rhode Island, a tribe of Bat People, the nocturnal shape-shifting vampire witches of Tlaxcala, and the immortal ways of Comte St. Germain in New Orleans and more. Softcover (200 pp/ 5.06" X 7.8") Suggested Retail: $12.99 ISBN: 978-1-953221-38-4

Mummies of the Americas explores Death Valley's city of the Dead, King Tut's Tomb along the Arkansas, the Egyptian City of the Grand Canyon plus the famous mummies of John Wilkes Boothe, Elmer McCurdy, the Cardiff Giant, the Mummy of Helldorado, and even Billy the Kid's pickled trigger finger! Softcover (200 pp/5.06" X 7.8") Suggested Retail: $12.99 ISBN: 978-1-953221-37-7

Cowboys & Dogmen is devoted to tales of werewolves of the Wild West including the dreaded Navajo skinwalker, the Watrous Werewolf, the Beast of the Land Between Lakes, the Hellhounds of El Dorado Canyon, the dreaded Dog Eater, the Wahhoo, the Wolf Man of Versailles, the Michigan Dog-Man and more! Softcover (212 pp/5.06" X 7.8") Suggested Retail: $12.99 ISBN: 978-1-953221-36-0

FICTION/ MISC. HISTORY

The first novel from historian John LeMay weaves a fantastic web of fiction via real life mysteries and legends of New Mexico, namely the puzzling theft and return of Billy the Kid's tombstone in 1976, the legend of the Lost Adams Diggings, the villainous Santa Fe Ring, and the enigmatic Acoma Mesa. Softcover (250 pp/5.5" X 7.5") Suggested Retail: $14.95 ISBN: 978-1-953221-42-1

The year is 1950, and old timers connected to the long-dead outlaw Billy the Kid are turning up murdered in New Mexico. Some blame the killings on the avenging witch of the Navajo nation, the skinwalker, while others think it's no coincidence that a man claiming to be a surviving Billy the Kid is set to meet with the governor soon... Softcover (260 pp/5.5" X 7.5") Suggested Retail: $16.95 ISBN: 978-1-953221-32-2

Roswell, USA, the long-forgotten debut work of LeMay, is available again and covers the minutia of the infamous Roswell UFO Crash of 1947. Notable chapters include tales of an alien ghost haunting the old airbase, monsters in the nearby Bottomless Lakes, and even a dinosaur sighting outside of town. Softcover (248 pp/6" X 9") Suggested Retail: $14.95 ISBN: 978-0-9817597-5-3

This biography, for the first time ever, tells the history of western journalist Ash Upson, who ghostwrote Pat Garrett's *The Authentic Life of Billy the Kid* in 1882 and also reproduces many of Upson's letters that detailed the harsh realities of frontier life in New Mexico during the turbulent Lincoln County War. Softcover (318 pp/5.5" X 8.5") Suggested Retail: $16.99 ISBN: 978-1953221919

ALSO AVAILABLE

NEW MEXICO'S LOST WORLDS & ENCHANTED LANDS

JOHN LEMAY

LEGEND & LORE OF THE
LOST ADAMS

JOHN LEMAY

From the author of *The New Mexico Book of Witches*

LA LLORONA

Her Kith & Kin

JOHN LEMAY

Tales of terror from the Southwest!

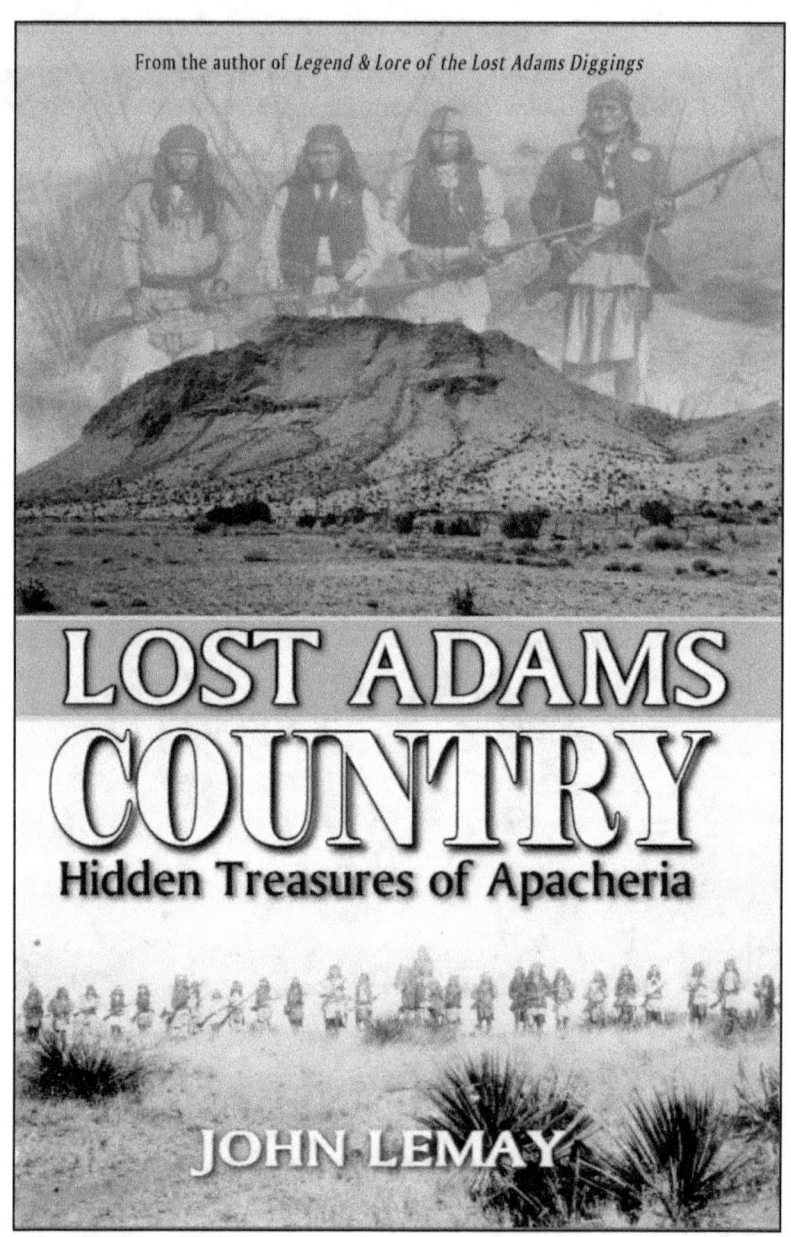

From the author of *Legend & Lore of the Lost Adams Diggings*

LOST ADAMS COUNTRY

Hidden Treasures of Apacheria

JOHN LEMAY

www.ingramcontent.com/pod-product-compliance
Lightning Source LLC
Chambersburg PA
CBHW060815120626
46557CB00001B/225